The NEC and You Perfect Together

RIVER PUBLISHERS SERIES IN POWER

Series Editors

MASSIMO MITOLO
Irvine Valley College
USA

The "River Publishers Series in Power" is a series of comprehensive academic and professional books focussing on the theory and applications behind power generation and distribution. The series features content on energy engineering, systems and development of electrical power, looking specifically at current technology and applications.

The series serves to be a reference for academics, researchers, managers, engineers, and other professionals in related matters with power generation and distribution.

Topics covered in the series include, but are not limited to:

- Power generation;
- Energy services;
- Electrical power systems;
- Photovoltaics;
- Power distribution systems;
- Energy distribution engineering;
- Smart grid;
- Transmission line development.

For a list of other books in this series, visit www.riverpublishers.com

The NEC and You Perfect Together:

A Comprehensive Study of the National Electrical Code

Gregory P. Bierals

Electrical Design Institute, USA

River Publishers

Routledge
Taylor & Francis Group

LONDON AND NEW YORK

Published 2021 by River Publishers

River Publishers

Alsbjergvej 10, 9260 Gistrup, Denmark

www.riverpublishers.com

Distributed exclusively by Routledge

4 Park Square, Milton Park, Abingdon, Oxon OX14 4RN

605 Third Avenue, New York, NY 10158

First published in paperback 2024

Library of Congress Cataloging-in-Publication Data

The NEC and You Perfect Together: A Comprehensive Study of the National Electrical Code / Gregory P. Bierals.

Routledge is an imprint of the Taylor & Francis Group, an informa business

Publisher's Note
The publisher has gone to great lengths to ensure the quality of this reprint but points out that some imperfections in the original copies may be apparent.

ISBN 978-87-7022-618-9 (online)

While every effort is made to provide dependable information, the publisher, authors, and editors cannot be held responsible for any errors or omissions.

NEC®, NFPA70®, NFPA70B®, NFPA70E® and National Electrical Code are trademarks of the National Fire Protection Association.

ISBN: 978-87-7022-619-6 (hbk)
ISBN: 978-87-7004-279-6 (pbk)
ISBN: 978-1-003-20731-3 (ebk)

DOI: 10.1201/9781003207313

Contents

Preface

It started on August 3, 1964 when I started to work in the electrical trade through the auspices of Local 52, IBEW in Newark, N.J. Seven months later, I became a member of Local 52, which was merged with Local 164 (Paramus, N.J.) in 2000.

I served my apprenticeship until March 15, 1969, at which time I became a Journeyman Wireman. I certainly gained a wealth of knowledge and experience during this period and the years that followed.

In March of 1978, I developed an association with a training/consulting company in Trenton, N.J. This company had a request from a client in Philadelphia to present an electrical training course for their maintenance personnel. This class was scheduled for four weeks. I was asked to conduct this course, and, despite not having any teaching experience, I decided to become an instructor. Fortunately, the class went well, and the training was extended for another four weeks. In the meantime, this training/consulting company offered me a permanent position.

In September of 1978, I started to offer courses on the topic of the National Electrical Code. The key was to find a method of instruction that would benefit the students by keeping their attention during the three day course period. And, to foster an interest in this complex document that would serve them well beyond our brief time together. In later years, I developed and presented courses on the topics of Grounding Electrical Distribution Systems, Designing Overcurrent Protection, Electrical Systems In Hazardous (Classified) Locations, and Electrical Equipment Maintenance. These courses were offered by my company, Electrical Design Institute, and several universities, including the University of Wisconsin, George Washington University, North Carolina State University, the University of Toledo, and the University of Alabama.

In 2021, I authored books entitled, The NEC and You, Perfect Together, Grounding Electrical Distribution Systems, and Designing Overcurrent Protection, NEC Article 240 and Beyond. These books are published by River Publishers.

Gregory P. Bierals
March 24, 2021

Article 90

Introduction

90.1 (A) – This Section defines the purpose of the NEC as a means to protect persons and property from the hazards associated with use of electricity.

The NEC is not intended to be used as a design specification, even though there is a great deal of design related information included throughout the text.

Also, the intent is that the Code is not to be used as an instruction manual for untrained persons.

90.1 (B) – As far as the adequacy of the provisions of the NEC, the intent is to provide an installation that is essentially free from hazard, but may not be efficient or adequate for good service. And, it does not include provisions for the possible, or even likelihood, of the future expansion of electrical use.

90.2 (A) – This Section includes the information relating to exactly what is covered by the NEC. It is typical that the Code covers the Premises Wiring System, which extends from the Service Point to the farthest outlet of the distribution system. The Service Point is the point of connection between the facilities of the Serving Utility and the Premises Wiring

But, what about a case where there is no serving utility? This would be the case if a customer had their own power supply source(s), such as an onsite generator, wind turbine, or photovoltaic system. In these instances, Article 230 (Services) would not apply, as the supply conductors to buildings or other structures would be classified as Feeders, and Articles 215 and 225 would apply. In the instance where the supply source(s) was a combination of a utility supply and other energy sources, Article 705 would also apply.

Section 90.3 identifies the arrangement of the Code. The NEC is divided into the Introduction (Article 90), followed by nine chapters. Chapters 1 through 4 apply in a general sense, and Chapter 5, 6, and 7 apply to special occupancies (Chapter 5), special equipment (Chapter 6), and other special conditions (Chapter 7). So, information in Chapters 1-7 may be modified by information in Chapters 5, 6, and 7. For example, the Equipotential Bonding

1

requirements for Swimming Pools in Section 680.26, wiring of Emergency Systems in Section 700.10, or the Selective Coordination of overcurrent devices in an Emergency System in Section 700.32.

Chapter 8, which covers Communications Systems, is independent of Chapters 1-7, unless these requirements are specifically referenced in Chapter 8.

Chapter 9 addresses Tables and Examples.

The Annexes that follow Chapter 9 are for informational purposes only, and not a part of any Code requirement.

Section 90.4 deals with the Enforcement of the Code by the Authority Having Jurisdiction. This AHJ may be a statewide, countywide, or local authority that is responsible for enforcing the rules of this Code. It is common that the AHJ may modify or supplement the NEC. In addition, it is also common that the authority may enforce previous code cycles and not the latest version. See Informative Annex H, Administration and Enforcement.

Due to certain local conditions, or even special circumstances, the AHJ may permit alternate installation methods, where safety is not compromised.

However, I have seen certain requirements that are enforced by the AHJ that do not increase safety at all. This may be due to a misunderstanding of the purpose of a specific requirement. For example, Section 250.53(A)(3) discusses supplemental electrodes. This Section states that, if multiple rod, pipe, or plate electrodes are installed to meet the requirements of this Section, they shall not be less than 1.8m (6 feet) apart. So, the AHJ may require that the maximum spacing between these multiple electrodes is 1.8m (6 feet). This spacing is inadequate, and for this reason, I submitted a proposal to add an Informational Note to this Section to indicate that the paralleling efficiency of rods is increased by spacing them twice the length of the longest rod.

Section 90.5(A) identifies that mandatory rules that specifically require or prohibit certain actions are referenced by the terms 'shall' or 'shall not'. If a rule is permitted, but not required, 'shall be permitted' or shall not be required' is used (90.5(B). Informational Notes that follow certain sections serve as a means to refer the reader to other standards or other helpful information that may apply. These notes are not mandatory and are not enforceable (90.5(C)).

Section 90.6 applies to a formal interpretation procedure by the NFPA Regulations Governing Committee Projects. The procedures for Formal Interpretations of the NEC are outlined in Section 6 of the NFPA Regulations Governing the Development of NFPA Standards.

Section 90.9 discusses Units of Measurement. Due to the fact that national standards are being harmonized with international standards, the NEC has included the International System of Units (SI) first, and the national standard measurement follows in parentheses.

Commentary Table 90.1 offers examples and comparisons of conversions of U.S. Customary units in inches and feet to equivalent SI units. A conversion chart is included at the end of this book.

One of the most common conversions that I have used often in my international travels is the need to convert American Wire Gauge (AWG) sizes to metric wires sizes. Without a chart to make this conversion, the easiest method is to divide the circular mil area by a factor of 1973.53.

Article 100

Article 100 includes the definitions of terms that are found in at least two Articles of the NEC. These terms may be unique, or even peculiar in nature. And they specifically relate to appropriate applications in Chapters 1 through 8.

Other Articles contain defined terms which apply only to that particular Article. For example, Article 240 (Section 240.2), which defines 'Current-Limiting Overcurrent Protective Device', 'Supervised Industrial Installation', and 'Tap Conductors', Article 250 (Section 250.2), which defines the term 'Bonding Jumper, Supply-Side', Article 310 (Section 310.2), which defines 'Electrical Ducts and Thermal Resistivity'.

*Accessible (as applied to equipment)-

Equipment that is considered 'accessible' would not be installed in locked rooms or isolated by elevation. Section 404.8(A) applies to switches and circuit breakers used as switches. Generally, the measurement from the floor or working platform to the center of the switch or circuit breaker handle, in its highest position, is 2 meters (6 feet, 7 inches). However, busway switches, that are floor operable through the use of ropes, chains, or a hookstick, would be an exception at higher elevations (368.10 (A),(B),(C), (368.17(C)).

*Accessible (as applied to wiring methods)-

Wiring methods are considered accessible where they may be removed or exposed without damaging the building structure or finish.

Probably, the most common example would be the wiring behind the removable ceiling panels of a suspended ceiling. Or, a fixed ceiling with doors or panels so that physical entry is possible. Also, please refer to Section 300.22(C) for information on acceptable wiring methods in ceiling cavities used for air-handling purposes.

*Accessible, Readily (Readily Accessible)-

Capable of being reached quickly for operation, renewal, or inspections without requiring those to whom ready access is requisite to take actions such as to use tools (others than keys), to climb over or under, to remove obstacles, or to resort to portable ladders, and so forth.

The concept of something being 'readily accessible' does not preclude the use of a lock(s) to exclude unauthorized persons from operating equipment that may be deemed critical to the safety of people or to a specific process. Sections 230.70(A)(1) and 230.205(A) require the service disconnecting means to be 'readily accessible'. Providing a lock on these disconnects is a relatively common practice, and if keys are available to those persons for whom ready access is necessary, then the equipment is readily accessible. This would not be the case if a tool is required to gain access.

*Adjustable Speed Drive-

This equipment is used to change the speed of an electric motor by controlling the frequency and voltage of the power supply. Article 430-Part X.

*Adjustable Speed System-

A combination of an adjustable speed drive, its associated motor(s), and auxiliary equipment. Article 430-Part X.

*Ampacity-

This term and definition first appeared in the 1965 Code cycle. At that time, it simply read – 'the current-carrying capacity of a conductor expressed in amperes'. It now states, 'the maximum current, in amperes, that a conductor can carry continuously, under the conditions of use, without exceeding its temperature rating'. Please refer to Section 310.15(A)(3).

The four considerations relating to conductor operating temperature are:

1. Ambient temperature, which is a variable, and, therefore, the worst case, at any point along the conductor length must be considered.
2. The heating of the conductor caused by load current, which, at times, may be normal 60 cycle current, as well as harmonic current.
3. The means of transferring the generated heat into the surrounding medium. Thermal insulation, in close proximity to the conductor, may affect its ability to dissipate heat.
4. The effect of adjacent load-carrying conductors, such as those in the same raceway or cable assembly.

The ambient temperature correction factors are listed in Table 310.15(B)(2), (a) and Table 310.15(B)(2),(b). Where the ambient temperature is above or below 30°C (86°F), Table 310.15(B)(2),(a) applies, and when the ambient temperature, is above, or below 40°C (104°F), Table 310.15 (B)(2),(b) applies.

For proximity effects, or the effects of adjacent load-carrying conductors in a raceway or cable where their number exceeds 3, ampacity adjustment factors are identified in Table 310.15(B)(3),(a). It should be noted that the number of current-carrying conductors includes spare conductors intended for future use. In addition, in a 3-phase 4-wire, Wye-connected system, where 2-phases and the neutral of this system are used, the neutral carries approximately the same current as the line-to-neutral currents of the other conductors, and it is to be counted as a current-carrying conductor (310.15(B) (5),(b)). The neutral conductor always carries current in this 3-wire system.

To calculate the neutral current, where a 3-wire circuit, consisting of two phases and a neutral is supplied from a 3-phase, 4-wire, Wye connected system is identified in Example 1.

Also, on a 3-phase, 4-wire, Wye-connected system, where the major portion of the load is nonlinear, the neutral conductor is to be considered a current-carrying conductor. In this case, the harmonic currents in the phase conductors do not cancel, as they normally would in a 60 Hz system. These harmonic currents add in the neutral conductor, and, for this reason, the neutral conductor is considered to be a current-carrying conductor (310.15(B)(5)).

To calculate the neutral current, where a 3-wire circuit, consisting of 2 phases and a neutral is derived from a 3-phase, 4-wire, Wye connected system, or to calculate the neutral current in a 3-phase, 4-wire, Wye connected system, where there is an unblananced load on the 3-phase conductors, the following formulas may be used.

Example 1

$$IN\sqrt{\left(L1^2 + L2^2 + L3^2\right) - \left(L_1 \times L_2\right) + \left(L_2 \times L_3\right) + \left(L_1 \times L_3\right)}$$

L1 – 80 amperes
L2 – 80 amperes
L3 – 0 amperes

$IN \ \sqrt{80 \ amperes^2 + 80amperes^2 + 0amperes^2 \ - 6400 + 0 \ amperes + 0 \ amperes}$

IN –6400 + 6400 + 0 – 6400 + 0 + 0 = 6400

$\sqrt{6400} = 80\ amperes$

Neutral Current = 80 amperes

Example 2

L1 – 200 amperes
L2 – 175 amperes
L3 – 145 amperes

$$IN\sqrt{(L1^2 + L2^2 + L3^2 - \left(L_1 \times L_2\right) + \left(L_2 \times L_3\right) + \left(L_1 \times L_3\right)}$$

$$IN\ \sqrt{40,000 + 30,625 + 21,025 - 35,000 + 25,375 + 29,000}$$

IN – 91,650 – 89,375 = 2275

$\sqrt{2275} = 48\ amperes$

Neutral Current = 48 amperes

Example 3

Three phase, four wire, Wye connected system with two phases and the neutral tapped to supply a three-wire system (310.15(B) (3), (a).

$$\sqrt{(L1^2 + L2^2) - (L_1 \times L_2)}$$

$$\sqrt{50A^2 + 50A^2 - 50A \times 50A}$$

$$\sqrt{2500 + 2500\ -2500}$$

$$\sqrt{2500} \qquad\qquad = 50\ \text{Amperes}$$

Equipment grounding or bonding conductors, which only carry negligible capacitive-charging current, or possibly short-time ground-fault current, are not considered current-carrying conductors when considering the effects of adjacent current-carrying conductors (310.15(B)(6)).

This adjustment factor does not apply to conductors in raceways where the length is limited to 600mm (24in), or less. In this respect, due to the short length, there is an effective heat sink on each end of the raceway, and the heat from the internal conductors is easily transferred to the outside conductors (310.15(B)(3),(a),(2)).

In addition, where underground conductors enter or leave a trench and physical protection is provided in the form of rigid metal conduit, intermediate metal conduit, PVC conduit, or reinforced thermosetting resin conduit, and for raceways have a length not exceeding 3.05m, or 10 feet, this adjustment factor does not apply (310.15(B)(3),(a),(3)).

Also, this adjustment factor does not apply to Type AC and Type MC cable under the following conditions:

1. The cables have no outer jacket
2. Each cable has not more than 3 current carrying conductors
3. The conductors are No. 12 copper (3.31mm^2)
4. And there are not more than 20 current-carrying conductors installed without maintaining spacing between the cable assemblies (310.15(B)(3),(a),(4))

If there are more than 20 current-carrying conductors installed and these cables are longer than 24 inches (600mm), without maintaining spacing, a 60% ampacity adjustment factor must be applied to the conductor ampacity.

Due to the fact that cable trays are not classified as raceways, Section 392.80 applies.

So, the adjustment factors of 310.15(A)(3),(a) for more than 3 current-carrying conductors in uncovered cable trays do not apply to the total number of conductors in the cable tray. But, this adjustment factor does apply to the number of individual current-carrying conductors in each cable. This provision applies to the individual cables, and not to the number of conductors in the cable tray (392.80(A)(1)).

If the cable tray is continuously covered for more than 6 feet (1.8m) with solid unventilated covers, the normal ampacities of Table 310.15(B) (17) and Table 310.15(B)(19) are permitted to be reduced to 60% of the Table values.

And, if multiconductor cables are installed in a single layer, and they are spaced at least one cable diameter apart, the allowable conductor ampacities may be determined in accordance with Table (310.15(B)(2),(3)), Informative Annex (B). This Table addresses the ampacities of multiconductor cables with not more than 3 insulated conductors, 0-2000 volts, in free air, based on an ambient temperature of 40°C. (104°F). The cable types are TC (Tray Cable), MC (Metal-Clad), MI (Mineral-Insulated), UF (Underground Feeder and Branch Circuit Cable), and USEC (Underground Service-Entrance Cable).

For single conductor cables in cable trays, the provisions of 310.15(A)(2) apply. This means that, if more than one ampacity applies to the conductors of a circuit, the lowest ampacity will apply. For example, where single conductor 500 kcmil Type THW copper cables (2000 volts, or less) are installed in ladder or ventilated trough cable trays, the normal ampacity of these cables is based on 392.22(B)(1),(b). that is, where the single conductors (500 kcmil copper) are installed in a single layer in uncovered trays, in accordance with the cable tray width and a reference to Table 392.22(B)(1), the conductor ampacities are based on 65% of the allowable ampacities in Table 310.15(B)(17), (392.80)(A)(2),(b).

The allowable ampacity of a 500 kcmil THW copper conductor from this Table is 620 amperes. This is compared to the same conductor size in a raceway at 380 amperes in Table 310.15(B)(16), with no more than 3 current-carrying conductors within the raceway.

If the ambient temperature is 45°C., the provisions of Table 310.15(B) (2),(a) would apply, and referring to Table 310.15(B)(2),(a), at 45°C. ambient temperature, the ampacity correction factor is 0.82.

The corrected ampacity is:
500 kcmil copper – 620 amperes (Table 310.15(B)(17)

$$\frac{620 \; amperes \times 0.65}{403 \; amperes} \qquad \frac{403 \; amperes \times 0.82}{330.46 \; amperes}$$

$$\frac{.65 \times 0.82}{0.553} \qquad \frac{620 \; amperes \times 0.533}{330.46 \; amperes}$$

This ampere rating, as compared to the same conductor size, where no more than 3 current-carrying conductors are installed in a raceway (conduit), at the same 45° C. ambient temperature, shows the advantage of using uncovered cable tray.
500 kcmil copper – 380 amperes (Table 310.15(B)(16)

$$\frac{\times 0.82}{312 \; amperes}$$

And finally, if both the ambient temperature correction factor and the ampacity adjustment factor for more than 3 current-carrying conductors apply, multiply both together to get the single equivalent factor.

For wires in raceways or cables that may be exposed to sunlight on rooftops, there must be at least 23mm (7/8in), measured from the bottom of the raceway or cable to the rooftop. If this measurement is less then this dimension, a temperature adder of 33°C (60°F) must be added to the outside temperature to determine the appropriate ambient temperature for the application of the correction factors in Tables 310.15(B)(2),(a) and 310.15(B) (2),(b). Type XHHW-2 insulated conductors are not subject to this ampacity adjustment. Additional information for ambient temperatures in various locations can be found in ASHRAE Handbook-Fundamentals.

*Arc-Fault Circuit Interrupter-

The first reference relating to the use of Arc-Fault Circuit Interrupters appeared in the 1999 NEC cycle, but the required use of this equipment did not take effect until the 2002 NEC cycle (210.12).

There were reports, before 1999, of fires in dwelling occupancies, hotels, motels, and dormitories caused by damage to items as simple as a 2-wire lamp cord. It is quite common to have a lamp(s) on a night stand in a bedroom and, over time, with the lamp cord behind the night stand, or table, and, with this furniture pressed against the wall, the lamp cord may be crushed, producing an arcing short-circuit (line-to-neutral). This arcing fault will typically not be cleared by the branch-circuit overcurrent device, which for a 15 or 20 ampere circuit breaker, may take 25–1000 amperes to promptly clear. This arcing fault, in the presence of combustibles, is the perfect recipe for a fire, and certainly, the reason why these devices are necessary.

A listed combination AFCI circuit breaker is referenced in 210.12(A)(1) as a means of protection against arcing-faults for dwelling units. And, 210.12(A) requires this protection for 15 and 20 ampere branch circuits that supply outlets or devices in kitchens, family rooms, dining rooms, living rooms, parlors, libraries, dens, bedrooms, sunrooms, recreational rooms, closets, hallways, laundry areas, or similar locations.

The listed combination Arc-Fault Circuit Interrupter provides protection from an arcing-fault between the ungrounded conductor and neutral, or series arcing, that is, arcing along a single conductor caused by a broken, frayed, or partially disconnected wire, producing a condition where the current arcs through the air-gap.

A dual-function Arc-Fault Circuit Interrupter and Ground-Fault Circuit Interrupter is designed to protect against both arcing-faults and ground-faults (210.8), where the ground-fault current exceeds 6 milliamperes.

Some Arc-Fault Circuit Interrupters are also designed to provide surge suppression (SPD).

Outlet type Arc-Fault Circuit Interrupters are available, and these types may be installed as the first outlet on the branch-circuit in accordance with 210.12(A)(4).

Where modifications or extensions are made to existing branch-circuits, a listed combination AFCI circuit breaker may be used, or a listed outlet type AFCI, located at the first receptacle outlet of the branch-circuit (210.12(D)).

However, in order for the AFCI to function properly, the circuit neutral conductors must be separated from other circuits, and not connected to the neutrals of other circuits downstream of the panelboard. For this reason, and especially in older installations, where these neutral connections may have been made, a combination AFCI outlet may be a better choice, where circuit extensions or modifications are to be made.

Arc Fault Circuit Interrupters are listed in accordance with UL 1699. They provide protection from arcing fault conditions and open the circuit if an arc-fault is detected. Also, please refer to Sections 210.12(A), 406.4(D) (4), 440.65, 550.25, and 690.11.

The AFCI required by 690.11 is for the direct current circuits, operating at 80 volts or higher between any two conductors of solar photovoltaic circuits. This device must be listed and identified for DC, and, specifically for Solar PV Systems.

*Associated Apparatus Hazardous (Classified) Locations-

Apparatus in which the circuits are not necessarily intrinsically-safe themselves, but that affects the energy in the intrinsically-safe circuits and is relied on to maintain intrinsic-safety. Such apparatus is one of the following:

1. Electrical apparatus that has an alternative type of protection for use in the appropriate Hazardous (Classified) Location. This alternative type of protection includes 'Explosionproof', 'Purged and Pressurized Enclosures or Vessels', 'Dust-Ignition Proof', etc.
2. Electrical apparatus not so protected that shall not be used within a Hazardous (Classified) Location.

An intrinsic-safety barrier is an example of 'Associated Apparatus'. This equipment is designed to limit the ignition capable energy, under normal and <u>abnormal</u> conditions, to a level that is below the ignition temperature of the specific gas, vapor, or dust in the environment.

Section 504.4 requires Associated Apparatus to be listed, and 504.10(B) permits the installation of Associated Apparatus in any Hazardous (Classified) Location for which it has been identified.

*Associated Nonincendive Field Wiring Apparatus Hazardous (Classified) Locations-

Apparatus in which the circuits are not necessarily nonincendive themselves, but affect the energy in nonincendive field wiring circuits and are relied upon to maintain nonincendive energy levels. Such apparatus is one of the following:

1. Electrical apparatus that has an alternative type of protection for use in the appropriate Hazardous (Classified) Location, such as explosionproof enclosures.
2. Electrical apparatus not so protected that shall not be used in a Hazardous (Classified) Location.

Nonincendive circuits are permitted for equipment in Class I, Division 2, Class II, Division 2, or Class III, Division 1 or 2 Locations (500.7(F)). This protection technique also applies to nonincendive equipment (500.7(G)) and to nonincendive components (500.7(H)).

A Nonincendive Circuit is defined in Article 100 as a circuit, in which any arc or thermal effect produced under intended operating conditions of the equipment, is not capable, under specified test conditions, of igniting the flammable gas/air, vapor/air, or dust/air mixture (ANSI/ISA12.12.01-2013).

*Authority Having Jurisdiction (AHJ)-

This is the individual or organization responsible for accepting equipment or materials, or installation methods and procedures, as required by this Code. Depending on location, the AHJ may be a federal, state, or local group or individual.

Certainly, the responsibilities of the AHJ are far reaching and this term is directly connected with the term 'Approved', which is defined as, 'acceptable to the Authority Having Jurisdiction'. The approval may be associated with the listing of a product or material. Or, the acceptance by the AHJ of equipment or materials that have been 'field evaluated' by a 'Field Evaluation Body'. In addition, the AHJ may approve an installation method, not specifically identified in the NEC.

See Informative Annex H-Administration and Enforcement.

*Bathroom-

An area including a basin, with one or more of the following- a toilet, a urinal, a tub, a shower, a bidet, or similar plumbing fixtures.

Since 1971, the NEC has required ground-fault circuit interrupter protection for 125 volt, single phase, 15 and 20 ampere receptacle outlets, in bathrooms of dwelling units. Many years ago, builders would divide the bathroom into 2 separate rooms, with the toilet in one room, and the basin (sink) in another, thereby eliminating the GFCI protection. It makes no difference if the plumbing fixtures are separated by a partition or wall, because the term 'bathroom' applies to the whole area (210.8(A)).

For other than dwellings, all single-phase receptacle outlets that are rated at up to 150 volts-to-ground or less, and up to 50 amperes, as well as 3-phase receptacles rated up to 150 volts-to-ground and up to 100 amperes, must have GFCI protection (210.8(B)(1)).

Section 210.52(D) requires at least one receptacle outlet, within 3 feet (900mm) of the outside edge of each basin (sink) in a bathroom.

In the 2020 NEC cycle, a change in 210.11(C)(3) requires the 20 ampere branch circuit for dwelling unit bathroom receptacles to supply only countertop receptacles, which are required to be within 3 feet of the sink. Other bathroom receptacles may not be supplied on this circuit.

This 20 ampere branch circuit may supply more than one bathroom for the countertop receptacles.

The Exception to this Section permits the single 20 ampere branch circuit to supply other than countertop receptacles where only one bathroom is supplied.

*Mobile Homes, Manufactured Homes, and Mobile Home Parks-

Section 550.12(E) requires the bathroom receptacle outlet(s) to be supplied by at least one 20-ampere branch circuit. And, this branch circuit cannot supply other outlets, except a receptacle that is part of a luminaire. The receptacle outlet(s) must have GFCI protection (550.13(B)).

*Recreational Vehicles and Recreational Vehicle Parks -

Section 551.41(C)(1) requires GFCI protection for each 125 volt, single-phase, 15 or 20 ampere receptacle outlet that is adjacent to a bathroom lavatory. The receptacle outlet may be installed on the side of the lavatory cabinet.

Also, GFCI protection is required for receptacle outlets in areas that are occupied by a toilet, shower, or tub in a Recreational Vehicle, (551.41(C)(3)).

*Battery System-

This defined term has a connection to Article 706 'Energy Storage Systems'. The definition states that a battery system is an interconnected battery subsystem(s), consisting of one or more storage batteries and battery chargers, and can include inverters, converters, such as a DC to DC converter that supplies the source circuit for DC utilization equipment, or the inverter for a Solar PV System.

Wind Power Systems and Solar PV systems cannot always be relied upon to generate power when their resource (wind and sun) is unavailable. So, an energy storage system may be used to improve the reliability of stand-alone systems at times when wind and sunlight are not available.

*Bonded (Bonding)-

Establishing electrical continuity and conductivity, either by mechanical means (locknuts, set screws, etc.,) or by a more assured means, such as a wire.

At one time, this term was defined as 'the permanent joining of metallic parts to form an electrically conductive path which will assure electrical continuity and the capacity to safely conduct any fault-current likely to be imposed'.

In any event, the bonding connection, whether this is through a fitting, such as a locknut, a bonding screw, or by a more assured means, such as a wire, must be able to establish continuity and have the current-carrying capacity to safely conduct the calculated fault current, until this current has been removed by the operation of the appropriate overcurrent device. Or, to assure continuity in order to eliminate the hazard of arcing at flexible raceways, boxes, and enclosures in Class I, Divisions 1 and 2 Hazardous (Classified) Locations (501.30).

And now, we examine a group of defined terms relating to the concept of 'Bonding', and their relationship to Article 250, as well as to other Articles, such as, the panelboard bonding requirements of 517.14. For example, the equipment grounding systems of the normal and essential branch-circuit panelboards must be bonded through the use of an insulated continuous copper conductor not smaller than No. 10 AWG. This bonding means is applicable, whether these systems are supplied from the same source, or from different supply systems that serve the patient care area of a Health Care Facility.

In addition, the 'Equipotential Bonding' requirements of Section 680.26 relate to the importance of proper bonding methods for swimming pools. It is vitally important to establish an equipotential plane in the vicinity of a swimming pool, where even small differences of potential could present a significant problem to persons within the pool, or even walking on a concrete deck surrounding the pool.

Another example of the equipotential voltage gradient is referenced in IEEE Standard 80-2000. This book addresses the grounding and bonding conditions in a substation or switchyard, where an equipotential bonding grid is established through a set of buried conductors, with all metallic frames of equipment, metal supporting structures, and surrounding metal fencing bonded to this grid. Section 250.194 covers the bonding and grounding of fences and metal structures surrounding or within a substation. The intention of these requirements is for the purpose of limiting the effects of touch, step, and transferred potential within, and approaching a substation environment. Where the surrounding metal fence is bonded to the ground mat of the substation at each corner and at intervals not in excess of 160 feet (50 meters), the fence fabric actually adds to the cross-sectional area of the ground mat. And where the ground mat extends a distance of 3 feet (1 meter) beyond the periphery of the fence, step and touch potential protection is provided for persons approaching the fence from the outside (250.194(A),(B)).

Any portion of the metal fence that may extend to a remote area may be interrupted with a nonmetallic portion of fencing to eliminate the possibility of transferring potential beyond the substation. Just think of the metal fence surrounding the substation as the core of a transformer. The collapsing magnetic field from the electrical lines within the switchyard will cut through the metal fence and induce a circulating current through the fence fabric. This circulating current may be significant, and without a proper grounding and bonding system, the resultant voltage-rise on the metal fence could produce a lethal touch potential.

IEEE Standard 80 recommends that large transformers have two bonding conductors, connected on opposite sides of this equipment to the ground mat.

*Bonding Conductor or Jumper-

These conductors may be relatively short, or, in some cases quite long, such as where several large conduits are connected by a wire through the use of bonding bushings (250.92(B)), (250.97), and (250.98).

*Bonding Jumper Equipment-

A bonding jumper between two or more portions of the equipment grounding conductor. This jumper (wire) is typically used where electrical continuity may be compromised, such as at concentric or eccentric knockouts, flexible metal conduit, where conductors are protected at over 20 amperes, in lengths of over 1.8 meters or 6 feet, or where the trade size exceeds 1¼″ (metric designator 35) (250.118,(5),(b),(c),(d)). Or liquidtight flexible metal conduit, where conductors are protected at more than 20 amperes, or, for metric designators 21-35, or trade sizes $^3/_4$″-1¼″, or where conductors are protected at over 60 amperes, or in lengths of over 1.8 meters or 6 feet (250.118(6),(b),(c),(d)).The bonding jumper size is based on the rating of the circuit overcurrent device and 250.122.

*Bonding Jumper, Main-

This bonding connection must be able to safely carry the ground-fault current from the equipment grounding system to the grounded conductor (the neutral conductor, except for corner-grounded Delta systems). From this point, the ground-fault current will return to the source, and when this circuit is complete, the overcurrent device protecting the circuit where the ground-fault has occurred, is supposed to promptly open or clear. If the system is properly designed and installed, the ground-fault current that passes through the main bonding jumper may be significant, but of a very short duration. This is the reason that the cross-sectional area of the main bonding jumper is smaller than the ungrounded conductors that supply the service. Table 250.102(C)(1)identifies these wire sizes, and Notes 1 and 2 of this Table address modifications to the Table sizes.

For example, if the service conductors are 2/0 or 3/0 copper (67.44mm²)–(85mm²), the main bonding jumper may be 4 AWG copper. In this case, the main bonding jumper has a cross-sectional area of 0.314 (2/0), or 0.249 (3/0) of the area of the service conductors.

$$\frac{4\ AWG\quad 41,740\quad circular\ mils}{2/0\quad 133,100\quad circular\ mils} = 0.314$$

$$\frac{4\ AWG\quad 41,740\quad circular\ mils}{3/0\quad 167,800\quad circular\ mils} = 0.249$$

Note 1 of Table 250.102(C)(1) states that if the ungrounded conductors that supply the service or extend from a separately-derived system are larger that 1100 kcmil copper or 1750 kcmil aluminum, the main bonding jumper or system bonding jumper must have an area of at least 12½ percent of the ungrounded conductors.

Example

Service Conductors 2000 kcmil − Copper × 0.125 = 250 kcmil-Copper

$$\frac{250,000\quad circular\ mils}{2,000,000\quad circular\ mils} = 0.125$$

In this case, the main or system bonding jumper size is 12½% of the ungrounded conductor size, or 250 kcmil copper (127mm²).

So, we are not concerned with the continuous current rating of this conductor, but the short-time current rating.

We will consider the following example of a three-phase system, with a Delta connected primary at 13,800 volts, and a Wye connected secondary at 480/277 volts. The transformer is owned by the serving utility and its size is 1000 kVA, the transformer marked impedance is 3.5%, and the transformer is located outside of the building or structure, and the major portion of the load is not nonlinear.

Transformer - 1000 kVA (Utility Owned)
Primary voltage – 13,800 volts (Delta)
Secondary voltage – 480/277 Volts (Wye)

Percent Impedance – 3.5%–Definition

The percent of the primary voltage applied to the primary windings of the transformer that will allow full-load secondary current to flow in the secondary windings, with the secondary leads short-circuited.

It should be noted that UL 1561 states that listed transformers, 25kVA and larger, have a percent impedance that may be 10% higher or lower than the marked value. Therefore, for worst case conditions, multiply the marked percent impedance by 0.9.

Transformers constructed to ANSI Standards have a plus or minus 7.5% impedance tolerance. However, we will use –10% in this example.

Full-Load Primary Current:

$$\frac{1,000,000\ volt - amperes}{13,800\ volts \times 1.732} = 41.84\ amperes$$

Full–Load Secondary Current:

$$\frac{1,000,000\ volt - amperes}{480\ volts \times 1.732} = 1203\ amperes$$

The length of the secondary conductors, from the transformer secondary terminals (bushings) to the service overcurrent device (in this case, current-limiting fuses) is 100 feet (30.48m).

The secondary conductors consist of 3 sets of 600 kcmil copper conductors in parallel (310.10(H)), 3 per phase, with a full-size neutral, or 1800 kcmil (912mm²), per phase and neutral.

The service overcurrent devices within the service equipment are 3-1200 ampere, current-limiting fuses. These fuses will operate and clear a short-circuit fault in less than ½ cycle (.008 seconds), when operating within their interrupting range. They will have an interrupting rating of 200,000 amperes, symmetrical. At least one manufacturer has current-limiting fuses with an interrupting rating of 300,000 amperes, symmetrical.

Note: These fuses will have a maximum rating of 1200 amperes because of 240.4(B)(3). The use of the next standard size of overcurrent device is not permitted above 800 amperes.

Based on these conditions, we are able to calculate the available short-circuit current at the transformer secondary and at the line terminals of the 1200 ampere current-limiting fuses in the service equipment.

We have established that the transformer full-load secondary current is 1203 amperes, with a marked percent impedance of 3.5%. Available short-circuit current at the transformer secondary is as follows:

$$\frac{100}{3.15(3.5 \times 0.9)} = 31.75$$

$$\begin{array}{r} 1203\ amperes \\ \times 31.75 \\ \hline 38,195\ amperes \end{array}$$

This is the short-circuit current at the transformer secondary.

However, this does not include a motor contribution. Due to the rotational energy of a motor load during short-circuit or ground-fault conditions, there will be a significant increase in fault current.

There will be an earth (ground) connection at the utility owned transformer from the neutral bushing.

If the entire load supplied by this transformer consisted of motors, simply multiply the motor load by 4, and add this to the calculated short-circuit current. In this case, 1203 amperes × 4 = 4812 amperes.

$$38,195\,amperes$$
$$\underline{4,812\,amperes}$$
$$43,007\,amperes$$

If the motor load was 25% of the total load, the calculation would be:

$$1203\,amperes$$
$$\underline{\times\,4}$$
$$4812\,amperes$$
$$\underline{\times.25}$$
$$1203\,amperes$$

$$38,195\,amperes$$
$$\underline{+\;1203\,amperes}$$
$$39,398\,amperes$$

In this example, we will not include a motor contribution.

We have extended a parallel set of 1800 kcmil copper phase conductors, with a full-size neutral, for a length of 100 feet (30.48m) to the service disconnect, which has 3-1200 ampere current-limiting fuses. In this example, the service conductors are installed in <u>nonmagnetic</u> raceways.

$$\frac{1.732\times100\,feet\times38,195\,amperes}{28033\times3\;(conductors\,per\,phase)\times480\,volts} = \frac{6,615,374}{40,367,520} = 0.1639$$

Note: 28033 is the "C" value of the service conductors in a nonmagnetic raceway

These values are equal to one over the impedance per foot based on resistance and impedance values in accordance with IEEE STD. 241 and IEEE STD. 242.

$$\frac{1}{1+0.1639} = 0.8592$$

$$38,195\;amperes\;(transformer\,secondary)$$
$$\underline{\times0.8592}$$
$$32,817\;amperes\,(at\,service\,disconnect)$$

The available short-circuit current at the service disconnect is 32,817 amperes.
    If the service conductors were installed in steel (magnetic) raceways, the available short-circuit current would be less, due to the magnetic properties of this raceway (conduit). The counter-electromotive force (CEMF), caused by the collapsing magnetic field of the alternating current (inductive reactance), will reduce the available short-circuit current at the service equipment.

Example

$$\frac{1.732 \times 100 \times 38,195}{22,965 \times 3 \times 480\,Volts} = \frac{6,615,374}{33,069,600} = 0.2000$$

$$\frac{1}{1 + 0.2000} = 0.8333$$

38,195 *amperes* (*at transformer secondary*)

$\times 0.8333$

31,828 *amperes* (*at service equipment*)

32,817 *amperes* (*nonmagnetic raceways*)

$-31,828$ *amperes* (*magnetic racweays*)

989 *amperes*

So, the short-circuit current that is available at the service equipment is reduced by 989 amperes with the use of the steel (magnetic) raceways.

These examples are based on single conductors in each of the 3 (paralleled) raceways.
    There will be a slight difference if the installation consisted of cable assemblies within these raceways.
    For the cable assemblies within nonmagnetic raceways, the results are as follows:

$$\frac{1.732 \times 100 \times 38,195}{32,154 \times 3 \times 480\,volts} = \frac{6,615,374}{46,301,760} = 0.1429$$

$$\frac{1}{1 + 0.1429} = 0.875$$

38,195 *amperes* (*at transformer secondary*)

$\times 0.875$

33,421 *amperes* (*at service equipment*)

For the cable assemblies within steel (magnetic) raceways, the results are as follows:

$$\frac{1.732 \times 100 \times 38,195}{28,752 \times 3 \times 480 \ volts} = \frac{6,615,374}{41,402,880} = 0.1598$$

$$\frac{1}{1 + 0.1598} = 0.8622$$

$$38,195 \ amperes \ (at \ transformer \ secondary)$$
$$\times 0.8622$$
$$\overline{32,932 \ amperes \left(at \ service \ equipment\right)}$$

$$33,421 \ amperes \ (nonmagnetic \ raceways)$$
$$\underline{-32,932 \ amperes \left(magnetic \ racweays\right)}$$
$$489 \ amperes$$

We began this analysis by calculating the short-circuit current of a 3-phase, 1000 kVA, 13,800 volt, Delta connected primary, and a 480/277 volt, Wye connected secondary transformer, at the transformer secondary, and at the service disconnect, where 100 feet (30.48m) of paralleled conductors (3 conductors, per phase), and a full size neutral were extended in nonmagnetic raceways, and, then, in steel (magnetic) raceways. In addition, the short-circuit current was calculated using cable assemblies instead of single conductors for comparison.

The phase and neutral conductors are 1800kcmil, copper, 4 × 600 kcmil in each raceway, and 3 raceways, based on the full-load secondary current of 1203 amperes, per phase.

There was no adjustment factor applied, based on the effects of adjacent load-carrying conductors (proximity effects), because the major portion of the load is not nonlinear (310.15(B)(5),(c)), and the neutral conductors are not considered to be current-carrying conductors, (310.15(B)(3),(a)).

The interrupting rating (110.9) of the service disconnect and the service overcurrent devices must equal or exceed the calculated short-circuit current. In our summary, we stated that the service overcurrent devices were 1200 ampere current-limiting fuses.

A current-limiting fuse is designed to interrupt short-circuit currents in less than ½ cycle (.008 seconds). Short-circuit currents reach a peak during the first ½ cycle. So, these devices will clear the short-circuit current before it reaches its most destructive peak.

The all-important 'Main Bonding Jumper' must be able to safely carry the fault-current, until the 1200 ampere current-limiting fuses operate. We have stated that the Main Bonding Jumper connects the equipment grounding system to the neutral (bus) within the service equipment. This is ground-fault current that occurs within, or downstream of the service equipment. And, in our example, the service equipment is downstream of the source (transformer).

The ground-fault will not normally be a bolted-fault, that is, where the ungrounded conductor makes contact with the equipment grounding conductor in such a way as to have these systems physically bolted together. More likely, and, in fact, the most common type of ground-fault is the arcing-type fault. In this case, the ground-fault will strike and restrike and the effective ground-fault current path may not return enough current to the source (transformer, generator, etc.) to promptly clear the circuit overcurrent protective device (if, at all). There will be a voltage-drop associated with the arcing fault and this compounds the problem, as the impedance of the ground-fault return path is markedly increased.

A bolted line-to-equipment ground-fault may produce a current flow of 25%–100% of a three-phase, <u>line-to-line bolted</u> fault, depending on how far from the source the ground-fault occurs. So, to use a ground-fault current equal to 50% of the short-circuit current produced by a three-phase bolted fault is certainly reasonable.

If we use our short-circuit current calculation, based on the paralleled 600 kcmil copper conductors in nonmagnetic raceways, which was determined to be 32,817 amperes, it is a safe assumption that a line-to-equipment ground-fault downstream of the service equipment may be 50% of the bolted three-phase short-circuit fault, of 32,817 amperes, or 16,409 amperes. And, if the ground-fault occurred within the service equipment, the Main Bonding Jumper would have to carry this ground-fault current until the service overcurrent device opens (1200 ampere fuses).

If we base the current-carrying capacity of the Main Bonding Jumper, which will be 250 kcmil copper (127mm^2), (12½% of 1800 kcmil-250.102(C)(1)) = 225 kcmil, or 250 kcmil), and, specifically, the ½ cycle (.008 seconds) ampere rating of this conductor in accordance with its ½ cycle fusing or melting current, we find that to be 386,050 amperes. So, the Main Bonding Jumper will easily carry the 16,409 amperes of ground-fault current without damage.

If the ground-fault developed on a feeder circuit or branch circuit downstream of the service equipment, the ground-fault current must be determine by calculation, as we have done for the transformer and service equipment. And, based on the operating characteristics of the feeder or branch-circuit overcurrent device, we may determine the proper overcurrent protection to assure that the purpose of the NEC (90.1(A)), that is, the practical safeguarding of persons and property from hazards arising from the use of electricity, is not compromised.

Keep in mind that the transformer secondary conductors (3-600kcmil copper conductors, in parallel, per phase, with a full-size neutral), have no overcurrent protection, except for the transformer primary overcurrent devices. Table 450.3(A) permits the primary overcurrent protection to be 600% of the primary current for a circuit breaker and 300% for a fuse, based on the transformer rated impedance of not more than 6%. The transformer rated impedance in our example is 3.5%.

$$\frac{1,000,000\,volt - amperes\,(1000kVA)}{13,800\,volts \times 1.732} = 41.838\,amperes$$

$$\begin{array}{ll} 41.838\,amperes & 41.838\,amperes \\ \underline{\times\;\;6\;\;(circuit\,breaker)} & \underline{\times\;\;3\;\;(fuse)} \\ 251\,amperes & 126\,amperes \end{array}$$

Table 450.3(A), Note 1(b) permits the use of the next commercially available rating or setting, where the calculated rating or setting does not correspond to a standard rating or setting.

The transformer primary magnetizing inrush current may be 25 times the full-load primary current for 0.01 seconds. So, a careful analysis of this inrush current and this effect on the primary overcurrent device(s) must be determined. Simply selecting the primary overcurrent protection from Table 450.3(A) will not always be acceptable.

However, this transformer is owned by the <u>serving electric utility</u>, so the NEC does not apply (90.2(B)(5)). The transformer primary overcurrent protection may be much higher than permitted by Table 450.3(A), and Note 1(b) of this Table.

It must be noted that in our example, if the transformer is owned by the customer and not the utility, the 1000 kVA transformer, with a primary voltage of 13,800 volts and the primary protection at, possibly, 251 amperes for a circuit breaker (or the next commercially available rating or setting), or a fuse at 126 amperes (or the next commercially available rating), 450.3 Informational Note No. 1 references 240.4, 240.21, 240.100, and 240.101. These Sections apply to overcurrent protection of the transformer secondary conductors.

For example, 240.21(B)(5) would apply to our example because the supply transformer is located outside of the building or structure and the secondary conductors terminate into a single set of fuses (1200 ampere fuses) that limit the load to the ampacity of the secondary (tap) conductors.

There is no limit to the length of these conductors, and in our example, they are 100 feet (30m) long. These conductors are provided with physical protection, as they are installed in a raceway, in compliance with 240.21(B)(1)

A fault on the secondary side of this transformer may have to be cleared by the fusing or melting of the faulted secondary conductor, and the fusing current of a 600kcmil copper conductor is extremely high. For one cycle (.016 seconds), it is 251,041 amperes.

And, due to the fact that this transformer is connected Delta-to-Wye, and there is a 30-degree phase shift, that is, the primary voltage leads the secondary voltage by 30 degrees, short-circuits and ground-faults on the secondary side of the transformer may not, or probably, will not, be cleared by the primary overcurrent devices.

*Available Short-Circuit Current -

Section 110.24(A) requires that the maximum available fault current be marked in the field on the Service Equipment. This requirement does not apply to dwelling units. The field marking must indicate the date that the calculation was made, and it must be of sufficient durability to withstand the environment (110.21(B)).

In addition, this calculation must be documented and made available to those people authorized to design, install, inspect, maintain, or operate the system.

Section 110.24(B) states that where modifications are made, and these changes affect the initial calculated fault-current, the available short-circuit current must be recalculated and the field markings on the service equipment must identify the new calculated fault current.

For service equipment rated 1200 amperes or more, a permanent label must be field or factory applied and include the following information (110.16(B)):

1. Nominal system voltage
2. Available fault-current at the service overcurrent device(s)
3. The clearing time of the service overcurrent devices, based on the available fault-current at the service equipment
4. The date the label was applied

Other equipment that must be identified with a short-circuit current rating are the following:

Meter disconnect switches up to 1000 volts (230.82(3))
Surge-Protective Devices (285.7)
Motor Control Centers (430.99)
Motor controllers of multimotor and combination load equipment for hermetic-sealed motor compressors and equipment (440.4(B)) (440.10(A),(B)).
Modular Data Centers (646.7)
Industrial Machinery (670.5)
Transfer Equipment (700.5(E))(701.5(D))
Direct-Current Microgrid Systems (712.72)

*Bonding Jumper, System-

This is the connection between the grounded (neutral) conductor and the equipment grounding conductor of a separately-derived system. This connection may be established at the source (transformer, generator, etc.), or at the first system disconnect or overcurrent device supplied from the source (250.28(A),(B),(C),(D))(250.30(A),(B)). No additional connections are to be made between these systems beyond the initial connection to prevent 'objectionable current' over the equipment grounding conductors (250.6).

And, just as in the case of the Main Bonding Jumper, the System Bonding Jumper may be significantly smaller than the ungrounded feeder conductors in recognition of the extremely high short-time current carrying capacity of these conductors (250.28(D)),(250.102(C)(1)).

*Branch Circuit-

The branch circuit conductors extend from the final overcurrent device protecting the circuit and the connected load. The rating of the branch circuit is established by the rating of this final overcurrent device, regardless of its size. There may be additional supplementary overcurrent devices in this circuit (240.10). However, the supplementary overcurrent device(s) may not be used as a substitute for the branch-circuit overcurrent device (UL 1077). Article 210 covers the general requirements for branch-circuits (210.1).

*Branch Circuit, General-Purpose-

This circuit supplies two or more outlets (210.23). Section 210.23(A) permits a 15 or 20 ampere branch circuit to supply lighting units or other utilization equipment. Or, combinations of both lighting units and utilization equipment. However, this does not include the small-appliance, laundry, and bathroom branch circuits as referenced in 210.11(C)(1),(2),(3). The circuit that supplies the receptacle(s) in these areas may not supply other receptacle outlets.

Section 210.23(A)(1)-addresses the rating of any one cord-and-plug connected utilization equipment which is not fastened in place. This utilization equipment may not have a rating that exceeds 80% of the branch circuit rating.

Section 210.23(A)(2)-covers the total rating of utilization equipment that is fastened in place. This equipment cannot exceed 50% of the branch-circuit rating. This does not include lighting units (luminaires). However, the luminaires, as well as cord-and-plug connected utilization equipment, may also be supplied on this branch-circuit.

*Branch Circuit, Individual-

This circuit supplies only one outlet. A single motor, heater, or range is supplied from an 'individual branch circuit' (210.21), (210.22). The general rule is that a single receptacle installed on an individual branch-circuit must have an ampere rating of not less than the rating of the branch-circuit (210.21(B)(1)).

The actual connected load on an individual branch-circuit may not exceed the ampere rating of the branch-circuit (210.22).

*Branch Circuit, Multiwire-

This is a single-phase or three-phase branch circuit with either two or three ungrounded conductors and a grounded conductor. There are equal voltages between the ungrounded conductors and the grounded conductor (210.4), (240.15(B)(1)), (300.13(B)).

It should be noted here that multiwire branch circuits are only permitted to supply line-to-neutral connected loads (210.4(C)). Exceptions permit multiwire branch circuits to supply line-to-line loads where only one piece of equipment is supplied, such an electric dryer (210.4(C), Exception No. 1), or where all of the ungrounded conductors are opened simultaneously by the branch-circuit overcurrent device, such as a common-trip circuit breaker (210.4(C), Exception No. 2). Where a multiwire branch circuit supplies only single-phase, line-to-neutral loads, identified handled-tied circuit breakers are permitted as the circuit protection (240.15(B)(1)).

For device removal, typically receptacle outlets, the continuity of the neutral is not to be affected by removal of the device. Splices of the neutral conductor, with a jumper wire to the device will allow the removal of the device, without affecting the continuity of the neutral conductor (300.13(B)).

Certainly, multiwire branch circuits offer some distinct advantages over individual branch circuits, especially in larger installations. These advantages include reduced voltage-drop, smaller conductors, and smaller raceways or cable assemblies.

However, the continuity of the neutral must be assured in order to prevent a serious hazard.

For example, consider a 240/120 volt multiwire branch circuit supplying two appliances. One appliance is rated 600 watts and the other is rated 1600 watts. If the neutral is disconnected, there will be an unequal voltage across each appliance.

$R = E^2/P = 14,400/600$ watts $= 24$ ohms
$R = E^2/P = 14,400/1600$ watts $= 9$ ohms
$I = E/R = 240$ volts $/(24$ohms $+ 9$ ohms$) = 7.3$ amperes
$E = I \times R = 7.3$ amperes $\times 24$ ohms $= 175$ volts
$E = I \times R = 7.3$ amperes $\times 9$ ohms $= 65$ volts
600 watts and 24 ohms $= 175$ volts
1600 watts and 9 ohms $= \underline{\ 65 \text{ volts}\ }$
$\qquad\qquad\qquad\qquad\quad 240$ volts

The overvoltage across the 600 watt appliance, due to the open neutral, will pose a significant fire hazard. This is the reason that 300.13(B) requires that the mechanical and electrical continuity of the grounded (neutral) conductor must be maintained and not affected by the removal of a device or other equipment.

Another potential hazard with the use of multiwire branch circuits is where the ungrounded conductors are terminated on the same line (single-phase), or the same phase (three-phase). In this case, instead of the neutral conductor carrying the maximum unbalanced load of the ungrounded conductors, the load current on the ungrounded conductors becomes additive in the neutral conductor, and this may, and probably will, cause this conductor to overheat and eventually fail.

For example, consider a three-phase, four-wire, multiwire branch circuit, protected by a 20 ampere 3-pole common-trip circuit breaker, and supplied with 12 AWG copper conductors, including the neutral conductor.

The phase conductors are connected to Phase A. The load on one ungrounded conductor is 15 amperes, on another it is 12 amperes, and on the third, it is 10 amperes. These currents would add in the neutral conductor, so the total load on this 12 AWG copper conductor is now 37 amperes. The general industry consensus is that the <u>average</u> service life of conductor insulation is <u>30 years</u>. However, where the insulation is subjected to a temperature of 5-10 degree C. above the insulation temperature rating, for extended periods, the insulation service life may be reduced by 50%.

Under these conditions, the neutral conductor will fail, and this may create a significant fire or safety hazard.

The termination of the neutral conductor will also be affected through heating and cooling cycles, and may eventually fail.

*Building-

A building is a structure which stands alone, or that is separated from adjoining structures by fire walls. This term has many applications relating to other parts of the Code, such as, grounding systems and equipment (250.20,21) (250.110), location of service equipment, (230.70), location of building disconnecting means (225.32), and requirements for fire stops (300.21).

The following references involve Outside Branch-Circuits and Feeders (Article 225), Services (Article 230), Grounding and Bonding (Article 250), and the General Requirements for Wiring Methods and Materials (Article 300). And they are relative to the definition of the term 'Building' in Article 100. A review of each Section listed here will further enhance your understanding of this term.

Article 225 - Outside Branch-Circuits and Feeders

Section 225.27 – Raceway seal
Section 225.30 – Number of supplies for buildings or structures
Section 225.31 – Disconnecting means for ungrounded conductors that supply buildings or structures
Section 225.32 – Location of disconnecting means
Section 225.33 – Maximum number of disconnects
Section 225.34 – Grouping of disconnects
Section 225.35 – Access to occupants (disconnecting means)
Section 225.36 – Type of disconnecting means
Section 225.37 – Identification of discounting means
Section 225.39 – Rating of disconnect
Section 225.52 – over 1000 volts – Location of disconnecting means
Section 225.56 – over 1000 volts – Pre-energization and operating tests
Section 225.60 – over 1000 volts – Clearances of conductors over roads, walkways, rails, water, open land

Section 225.61 – over 1000 volts – Clearances over buildings and other structures

Article 230 - Services

Section 230.2 – Number of services for a building or structure
Section 230.3 – Building or structure not to be supplied through another building or structure
Section 230.6 – Conductors considered outside of the building
Section 230.8 – Raceway seal
Section 230.9 – Clearances on buildings
Section 230.70– Service equipment – disconnecting means
Section 230.71 – Maximum number of disconnects
Section 230.72 – Grouping of disconnects
Section 230.82 – Equipment connected to the supply-side of service disconnect
Section 230.200 – Services exceeding 1000 volts
Section 230.204 – Isolating switches
Section 230.205 – Disconnecting means
Section 230.208 – Short-circuit protective device for ungrounded conductors over 1000 volts

Article 250 - Grounding

Section 250.4 – General requirements for grounding and bonding
Section 250.106 – Lightning protection systems
Section 250.30 – Grounding separately–derived systems
Section 250.32 – Buildings or structures supplied by a feeder(s) or branch circuit(s)
Section 250.58 – Common grounding electrode (system)
Section 250.64 – Grounding electrode conductor installation
Section 250.66 – Size of AC grounding electrode conductor
Section 250.68 – Grounding electrode conductor and bonding jumper connection to grounding electrodes
Section 250.90 – Bonding
Section 250.92 – Bonding of equipment for services
Section 250.94 – Bonding for communication systems
Section 250.104 – Bonding of piping systems and exposed structural metal
Article 300 – General Requirements for Wiring Methods and Materials

*Circuit Breaker-

This device may be operated manually, or, sometimes, remotely, to open or close a circuit, and automatically deenergize a circuit due to an overload or fault condition without damage.

Circuit breakers have an identified interrupting rating, if this rating exceeds 5000 amperes, (the <u>minimum</u> interrupting rating is now 10,000 amperes (240.83(C), (110.9)).

Circuit breakers have a marked voltage rating which is associated with the interrupting rating of the unit. Circuit breakers with a <u>straight</u> voltage marking, e.g., 480 volts, may be used on ungrounded systems (480 volts), or, on systems that are solidly grounded (480/277 volts), (240.85).

Circuit breakers with a slash voltage marking (120/240 volts), (480Y/277 volts) <u>may only be used in solidly grounded systems.</u> These devices are not intended to open phase-to-phase voltages across one pole, as may occur when one phase of a three-phase circuit develops a ground-fault. In recognition of this potential problem, Section 240.85 limits their use to solidly grounded systems.

It should be noted that motor controllers are marked with horsepower ratings, which indicate their ability to <u>interrupt</u> locked-rotor currents. (430.83(A)(1), (110.9)). And, the horsepower ratings must be at least equal to the horsepower rating of the motor at the application voltage.

And, Section 430.83(E) states that controllers with slash voltage markings (120/240 volts), (480Y/277 volts) are only suitable for systems that are solidly grounded. Whereas, those controllers with a straight voltage marking (480 volts) may be used on ungrounded systems, or on those systems that are solidly grounded.

Circuit breakers may be identified as 'Current-Limiting' (240.2). These devices are designed and tested to open under fault conditions, within ½ cycle (.008 seconds), or less (IEEE-1584-Table 1), when the fault-current is within the circuit breaker current-limiting range. A short-circuit current reaches a peak within the first ½ cycle. So, where the overcurrent protective device opens in the first ½ cycle, the short-circuit current is interrupted as it reaches its peak level. Better protection for circuit components is the result (110.10). However, selective coordination with downstream overcurrent devices is very important where these devices are not current-limiting, and may take a longer time to open. This may cause the current-limiting device to operate before the downstream noncurrent-limiting device(s), and cause unnecessary outages.

*Circuit breaker (Adjustable)-

This device can be set to trip at various values of current, time, or both. Section 240.6(A) identifies the standard ratings for fuses and inverse-time circuit breakers. Those devices with nonstandard ratings are also acceptable.

For adjustable-trip circuit breakers, the rating is considered to be the maximum possible setting (240.6(B)).

However, if the adjustable-trip circuit breaker is provided with limited access to its adjusting means, such as through the use of removable covers over the adjusting means, or bolted equipment doors, or locked doors, where access is permitted only to qualified persons, the circuit breaker rating is considered to be the setting of its adjusting means (240.6(C)).

*Circuit Breakers (Instantaneous-Trip)-

This device has no time-delay feature incorporated in its tripping action.

Instantaneous-trip circuit breakers that are used for the branch-circuit, short-circuit, and ground-fault protection of a motor must be of the adjustable type and part of a combination motor controller which has appropriate motor overload protection (430.52(C)(3)).

Table 430.52 identifies the maximum rating or setting of motor branch-circuit, short-circuit, and ground-fault protection. For instantaneous-trip circuit breakers, this protection may be set at 800% of the motor ampere rating, from Tables 430.248 (single-phase), 430.249 (two-phase), and 430.250 (three-phase), and not from the motor nameplate current rating (430.6(A)(1)).

Instantaneous-trip circuit breakers may be adjusted to 250% of the full-load current rating of a constant-voltage DC motor.

Motor overload protection is based on the motor nameplate current rating (430.6(A)(2)).

Section 430.52(C)(3), Exception No. 1 permits the setting of an instantaneous-trip circuit breaker to be increased from 800% to 1300% for other than Design B energy-efficient motors, or up to 1700% for Design B energy-efficient motors, where lower settings are not sufficient to accommodate the starting current of the motor.

Exception No. 2 applies to motors that have full-load current ratings of up to 8 amperes to be protected by an instantaneous-trip circuit breaker which has its adjustment set at 15 amperes or less, and is part of a combination controller with coordinated overload protection.

I would like to point out that the setting of the instantaneous-trip circuit breaker, as it relates to the starting current of the motor, is an important consideration. But, this circuit breaker is part of the entire <u>motor circuit protection scheme</u>. It must be capable of accommodating the starting current of the motor, so, the setting of its adjustment may be quite high. This device is the branch-circuit, short-circuit, and ground-fault protection for the entire circuit, which includes the circuit conductors, controller, overloads, disconnect, etc. In determining the setting of the circuit breaker to allow for the inrush current of the motor, it is also necessary to determine that short-circuit and ground-fault protection of the circuit have not been compromised (110.10).

The problem with the use of magnetic only or thermal magnetic circuit breakers on motor circuits may be the inability to handle motor starting currents without tripping, unless they are rated or set at the maximum ratings or settings permitted by Table 430.52. This would be 250% of the motor full-load current rating for the inverse-time circuit breaker or 800% for instantaneous-trip circuit breaker (1100% for Design B Energy efficient motors). Rating or settings this high may eliminate the short-circuit protection for the motor overload protective devices.

Example

*Motor Circuit-

An industrial installation includes the following motor circuit:

One 25 horsepower – 460 volt, 3-phase squirrel-cage motor, Design B, nameplate full-load current – 30 amperes, service factor – 1.15, temperature-rise-40°C.

One – 10 horsepower – 460 volt, 3 phase squirrel–cage motor, Design B, nameplate full–load current – 11 amperes, service factor – 1.15, temperature rise -40°C.

One – 20 horsepower – 460 volt, 3-phase wound-rotor motor, nameplate primary full-load current 35 amperes, nameplate secondary full-load current 65 amperes, 40°C temperature – rise.

Branch–circuit short-circuit and ground fault protection (time-delay fuse or inverse-time circuit breaker)(430.52).

Time-Delay Fuse

25 hp– squirrel-cage motor –175% of full-load current
10hp – squirrel – cage motor –175 % of full – load current
20hp – wound –rotor motor – 150% of full – load current

Inverse – Time Circuit Breaker

25hp – squirrel-cage motor –250% of full-load current
10hp – squirrel – cage motor–250 % of full– load current
20hp – wound –rotor motor –150 % of full – load current

Conductor Ampacity

430.22 (squirrel-cage motors)
430.23(A) (wound-rotor (secondary)
The full-load current is selected from Table 430.250 (430.6(A)(1)).
25 hp–squirrel-cage motor – 34 amperes x 1.25 = 43 amperes
10hp–squirrel–cage motors – 14 amperes x 1.25 = 18 amperes
20hp–wound–rotor motor – 27 amperes x 1.25 = 34 amperes

430.52

Time –Delay Fuse

25*Hp*	10*HP*	20*HP*
34 *amperes*	14 *amperes*	27 *amperes*
×1.75	×1.75	×1.50
59.5 *amperes*	24.5 *amperes*	40.5 *amperes*
60 *amperes*	25 *amperes*	45 *amperes*

430.52(C), Exception No. 1 permits the use of the next standard size of time-delay fuse.

Inverse-Time Circuit Breaker

25*Hp*	10*HP*	20*HP*
34 *amperes*	14 *amperes*	27 *amperes*
×2.5	×2.5	×1.5(430.52)
85 *amperes*	35 *amperes*	40.5 *amperes*
90 *amperes*	35 *amperes*	45 *amperes*

430.52(C), Exception No. 1 permits the use of the next standard size of inverse-time circuit breaker.

Motor Overland Protection

430.6(A)(2)

430.32

25 *Hp* (*nameplate*)	14 *HP* (*nameplate*)	20 *Hp* (*nameplate*)
30 *amperes*	11 *amperes*	35 *amperes*
×1.25	× 1.25	×1.25
37.5 *amperes*	13.75 *amperes*	43.75 *amperes*

Section 430.32(C) permits the next higher trip setting of the overload relay where the setting at 125% is not sufficient to start the motor and carry the load, providing the overload trip-current does not exceed the following: motors with a marked temperature-rise of 40°C, or less -140% all other motors – 130%

Most overload relays are Class 20, which afford a 20 second time-delay. A Class 10 overload relay has a 10 second time-delay, and a Class 30 overload relay has a 30 second time-delay. In order to provide the best protection from overheating, it may be better to increase the time delay feature, rather than to increase the setting of the overland relay.

Motor Feeder

Short-Circuit and Ground-Fault Protection

430.24
430.62
430.94

Motor feeder overcurrent device-<u>Inverse-Time Circuit Breaker</u>. The feeder size is based on 430.24(1),(2), that is, 125 % of the full-load current rating of the highest rated motor, from 430.6(A)(1), plus the sum of the full-load currents of the other motors.

25HP-460V-3-Phase	10HP-460V-3-Phase	20hp-460v-3-Phase
Design B – FLA – 34 Amps	Design B – FLA – 14 Amps	Wound –Rotor-FLA-27Amps
Table 430.250	Table 430.250	Table 430.250

430.24

$$\begin{array}{r} 34\,amperes \\ \times 1.25 \\ \hline 43\,amperes \end{array} \;(42.5\,A)(220.5(B)(Informative\,Annex\,D)$$

$$\begin{array}{r} 14\,amperes \\ +\;27\,amperes \\ \hline 84\;\;amperes \end{array}$$

No. 3 AWG-Copper -85 amperes – 60°C, - Table 310.15(B)(16) -110.14(C) (1)(a)

Motor Feeder-Short-Circuit and Ground-Fault Protection

430.62 (A)

Inverse–Time Circuit Breaker

$$25 \text{ HP- } 34 \text{ amperes } \times 2.5 = 85 \text{ amperes}$$

$$\begin{array}{r} 14 - 10\,HP \\ +\;27 - 20\,HP - \quad wound - rotor \\ \hline 126\,amperes \end{array}$$

The rating of the inverse-time circuit breaker, cannot exceed 126 amperes. The standard size from 240.6(A) is 125 amperes.

Time-Delay Fuse

$$\begin{array}{r} 25HP - 34\;\;amperes \\ \times\;1.75 \\ \hline 59.5\;\;amperes \end{array} \quad (Time - Delay\,Fuse - Table\,430.52)$$

$$\begin{array}{r} 14 \\ +\;27 \\ \hline 100.5 - 101\,amperes \end{array}$$

The rating of the Time-Delay fuse cannot exceed 101 amperes. The standard size from 240.6(A) is 100 amperes.

The Note on the bottom of Table 430.22(E) states that any motor application shall be considered as continuous-duty, unless the nature of the apparatus it drives is such that the motor will not operate continuously with load under any conditions of use.

Table 430.22(E) identifies four duty cycles other than continuous duty, and specific multipliers that are applied to the motor nameplate current rating in order to size the motor circuit conductors.

Examples of These Duty Cycles Are:-

Short-time-duty (operating valves, etc.), Intermittent-duty (freight and passenger elevators, etc), Periodic-duty (rolls, ore and coal-handling machines), Varying-duty (conveying machines).

Table 430.22(E) specifies the appropriate percentages that are applied to the motor nameplate current rating, depending on the time rating of the motor. The four time ratings are:

- 5 Minute Rated Motor
- 15 Minute Rated Motor
- 30-60 Minute Rated Motor
- Continuous Rated Motor

For example, a varying-duty–15 minute rated motor will have conductors sized in accordance with a multiplier of 1.2 (120%) applied to the motor nameplate current rating.

*Circuit Breaker-

Circuit breakers are mechanical overcurrent protective devices and they have a current sensing means that is thermal, magnetic, or electronic.

There is a mechanical unlatching mechanism and a current/voltage interrupting means, which includes the mechanical contact parting and a means of cooling and quenching an arcing condition in an arc chute.

A typical thermal magnetic circuit breaker has a bimetal element and a magnetic element. The bimetal element is sensitive to an overload condition. When the overload continues for an extended period, the bimetal element starts to bend and it exerts pressure on the circuit breaker trip bar, which unlatches the circuit breaker, and the opening of the circuit breaker spring-loaded contacts is irreversible.

During short-circuit conditions, the magnetic element attracts the circuit breaker trip bar, which unlatches the circuit breaker contacts and forces the arc into the arc chute. And when the arc is extinguished, the circuit is open.

A standard molded case circuit breaker (600 volts), with no short-time delay feature, has a fault clearing time of 0.025 seconds. (1.5 cycles – 60Hz) (IEEE-1584- Table 1).

In some circuit breakers, the tripping function is performed by electronic means.

*Circuit Breaker (Inverse-Time)-

As is the case with all overcurrent protective devices, as the current through the device increases, the clearing time decreases. Certain types of these devices, namely, time-delay fuses, or inverse-time circuit breakers, have a purposeful time vs. current characteristic. In the case of a time-delay fuse, the device is designed to carry five times its current rating for at least ten seconds.

The time-delay feature of these fuses and inverse-time circuit breakers make them suitable for electrical loads that have a high degree of inrush current, such as motors and transformers.

Table 430.52 recognizes that the rating of an inverse-time circuit breaker may be up to 250% of the motor full-load current rating (Tables 430.248, 430.249, or 430.250). Section 430.52(C)(1), Exception No. 1, permits the use of the next higher device rating where this calculation results in a value that does not correspond to a standard size (240.6(A). And Section 430.52(C)(1), Exception No. 2, permits the rating of an inverse-time circuit breaker to be increased from 250% to 400% for full-load currents of 100 amperes or less, or 300% for full-load currents of over 100 amperes.

For wound-rotor or DC (constant voltage) motors, the inverse-time circuit breaker may be rated at 150% of the motor full-load current rating.

*Clothes Closet-

A nonhabitable room or space intended primarily for the storage of garments or apparel. Section 240.24(D) prohibits the installation of overcurrent devices within a clothes closet, or where other easily ignitable materials may be present.

Section 410.16(A) recognizes the installation of surface-mounted or recessed incandescent or LED luminaires with completely enclosed light sources within a clothes closet. Surface-mounted or recessed fluorescent luminaires are also permitted.

In addition, surface-mounted fluorescent or LED luminaires that are specifically identified for installation within a clothes closet are acceptable.

Section 410.16(B) prohibits the installation of open or partially enclosed incandescent lamps or pendant luminaires within the closet storage space.

Section 410.16(C) identifies the minimum clearance between the luminaires installed in clothes closets and the nearest point of a closet storage space.

These clearances are:

12 inches (300 mm) for surface-mounted incandescent or LED luminaires with a completely enclosed light source on the wall above the door or on the ceiling.

6 inches (150 mm) for fluorescent luminaires that are surface-mounted on the wall above the door or on the ceiling.

6 inches (150 mm) for recessed incandescent or LED luminaires with completely enclosed light sources installed in the wall or ceiling.

6 inches (150 mm) for recessed fluorescent luminaires installed in the wall or ceiling.

Surface-mounted fluorescent or LED luminaires are permitted to be installed within the closet storage space where identified for this use.

Section 424.38(B)(1) prohibits the installation of 'Heating Cables' in a closet. However, Section 424.38(C) allows the installation of heating cable in a closet as a 'low temperature heat source' to control humidity where the cabling is not obstructed by shelving or permanent luminaires.

Section 550.11(A) requires the 'panelboard' for a Mobile or Manufactured Home to be in an accessible location, but not located in a bathroom or closet. This correlates with 240.24(D), which is applicable to any type of occupancy.

*Coaxial Cable-

A cylindrical cable assembly composed of a conductor and inside a metallic tube or shield, separated by a dielectric material and usually covered by an insulating jacket. The metallic tube or shield serves as the return conducting path. This type of cable has been in use for many years, along with twisted pair cable. However, these types of cables are being replaced with Optical Fiber Cables.

*Combustible Dust-

Dust particles that are 500 Microns or smaller present a fire or explosion hazard when dispersed and ignited in air (See U.S. No. 35 Standard Sieve as defined in ASTM E11-2015).

It should be noted that any type of dust, in the proper concentration and in a specific environment, is combustible.

Combustible dusts are subdivided into 3 specific groups. These are:

Group E–Combustible Dusts, including metallic dusts, such as aluminum and magnesium dust (500.6(B)). These dusts have relatively low ignition temperatures, in some cases 20°C., as well as low ignition energies.

In addition, metallic dusts produce higher explosion pressures than dusts that are more highly resistive (Groups F and G).

Group F– Includes semi conductive carbonaceous dusts, including coal, carbon black, charcoal, and coke dusts. While not as volatile as metallic dusts, when these types of dusts combine with moisture, and layers of dust form on heat producing equipment, the elevated temperature may cause the spontaneous ignition of the material.

Group G– This group contains highly resistive grain, flour, plaster, and wood dust. These dusts include flour, grain, wood, plastic, and chemicals. While these dusts are typically highly resistive and they do not present the same hazards as those of Groups E and F, from an ignition temperature and explosion pressure standpoint, they are quite volatile, especially when suspended in air in a thick concentration in the presence of even static charges. NFPA 499 is the 'Recommended Practice for the Classification of Combustible Dusts and of Hazardous (Classified) Locations for electrical installations in chemical process areas'.

Combustible dusts are classified and identified in NFPA 499-2013.

*Combustible Gas Detection System-

A protection technique utilizing stationary gas detectors in industrial establishments. This protection technique is restricted to those industrial establishments with restricted public access and where the conditions of maintenance and supervision ensure that only qualified persons service the installation (500.7(K)).

Where properly applied, the area classification of Class I, Division 1, due to inadequate ventilation, may use electrical equipment suitable for Class I, Division 2 when the combustible gas detection equipment is listed for Class I, Division 1, and the specific material group. In addition, where the interior of a building does not have a source of ignitible gases or vapors, and combustible gas detection equipment, listed for Class I, Division 1 or 2, and the appropriate group material is installed, equipment suitable for unclassified locations is permitted.

Where a control panel contains instruments utilizing or measuring flammable liquids, gases, or vapors, equipment suitable for Class I, Division 2 is acceptable for the appropriate group material encountered.

*Communications Equipment-

This equipment is designed to perform telecommunications functions in the transmission of audio, video, and data. It also includes the power systems, including batteries, converters, inverters, etc., and technical support equipment (computers), and the related conductors for the operation of the equipment. Also, see NFPA 76.

Article 800 covers 'Communication Systems' and Section 800.2 indentifies 'Definitions' that are applicable to Article 800. And, Chapter 8 is independent from Chapter 1–7, unless these requirements are also referenced in Chapter 8 (90.3).

Section 800.21 states that there shall not be an accumulation of communications wires and cables that block access to electrical equipment, such as the removal of panels that are designed to allow access. In this case, Section 800.21 coincides with Sections 725.21 (Class 1, Class 2, and Class 3 Circuits), 760.21 (Fire Alarm Circuits), 770.21 (Optical Fiber Cables), 820.21 (CATV Cables), 830.21 (Network-Powered Broadband Communications Circuits), and 840.21 (Premises-Powered Broadband Communications Systems).

The grounding and bonding methods for communications systems are more specific than the general information in Article 250. Part IV of Article 800 addresses these grounding and bonding methods.

For example, the minimum size of the grounding electrode conductor for the 'Primary Protector' is No. 14 AWG (2.083 mm^2), and is not required to be larger than No. 6 AWG (13.30mm^2) (800.100(A)(3). The length of this conductor is limited to 20 feet (6.0 m), and this conductor is to be run in a straight line, avoiding unnecessary bends (800.100(A)(4)(5). These provisions are for important impedance reduction.

The various types and identification of Communications Cable are listed in Section 800.154, depending on the application.

*Communications Raceway-

This is a nonmetallic enclosure expressly designed for communications wires and cables, optical fiber cables, and data cables associated with Information Technology and Communications Equipment, Class 2, Class 3,

Power-Limited Tray Cables, and Power-Limited Fire Alarm Cables for plenum, riser, or general-purpose use.

*Concealed-

Any wiring method that is rendered inaccessible by the structure or finish of the building is considered 'Concealed.' An 'Informational Note' indicates that wires in concealed raceways are considered concealed, even though they may become accessible by withdrawing them. This is opposed to the concept of 'Exposed' as applied to wiring methods, as defined in Article 100.

*Conduit Body-

This accessory 'fitting' is provided with a removable cover(s) in order to gain access to the interior of a conduit or tubing. (LB, LL, LR, LRL, C, T, or X designs). Conductor fill for the conduit body is typically the same as for the conduit or tubing (one conductor, or cable, 53%, two conductors, or cables, 31%, more than two conductors, or cables, 40%. (Table 1, Chapter 9). Only those conduit bodies that are durably and legibly marked with their volume in cubic inches may contain splices, taps, or devices in accordance with 314.16(B). Conduit bodies must be supported in a rigid and secure manner (314.23).

*Continuous Load-

A load where the maximum current is expected to continue for three hours or more. Section 210.20(A) identifies the conditions of continuous loading on branch circuits. Due to the increased heating effects caused by loads that operate continuously, the overcurrent device protecting the circuit is to be sized at 125% (1.25) of the continuous load, unless the overecurrent device(s) is part of an assembly that is listed to operate at 100% of its rating. This provision also applies to the size of branch-circuit conductors when supplying continuous loads (210.19(A)(1),(a)).

This requirement also applies to feeders that supply continuous loads (215.2(A)(1),(a)).

However, the concept of continuous loading does not apply to most motor applications. The definition of Continuous Duty from Article 100 is 'operation at a substantially constant load for an indefinitely long time'. This is the most common duty cycle for motors. Section 430.33 applies to the protection of the other duty cycles for motors. Generally, where short-time, intermittent, periodic, and varying duty cycles apply, overload protection is not required, as the motors are inherently protected by the motor branch-circuit overcurrent device in accordance with Section 430.52 (430.33). For conductor sizing for motor duty cycles other than continuous, see Table 430.22(E). For continuous duty, motors normally require overload protection in accordance with Sections (430.32(A)(1),(2),(3),(4), (430.32 (B), 430.32 (C), 430.32(D).

If a continuous-duty motor has overcurrent protection (fuse or circuit breaker) which does not exceed that required for overload protection, then a separate overload device is not required. For example, a motor with a marked service factor of 1.15, which normally requires overload protection at 125% of the motor nameplate current rating, and this motor is protected by a time-delay fuse rated at 125% of the motor nameplate current rating. Thermal/magnetic circuit breakers and magnetic only circuit breakers may not be able to handle motor starting currents at 115% or 125% of the motor nameplate current rating without tripping. A time-delay fuse is designed to carry current at 5 times its rating for 10 seconds. Keep in mind that the overload protection is based on the motor nameplate current rating, and not the appropriate Table ampere ratings in Article 430 that are used to size the motor circuit conductors and overcurrent protection, as well as motor disconnecting means (430.110).

Motor circuit conductors are based on 125% of the motor full-load current rating (430.22(A through G).

Example

Branch Circuit

210.19(A)(1),(a)

Continuous Load- 30 amperes
Noncontinuous Load– 15 amperes

Continuous Load	30 amperes
Noncontinuous Load	+ 15 amperes
Total	45 amperes

$30\,amperes$
$\times 1.25\ (210.19(A)(1)(a)$
$\overline{37.5, or\,38\,amperes}$ (220.5(B),(Informative Annex D)

$38\quad amperes\,(continuous)$
$+15\ amperes\ (noncontinuous)$
$\overline{53\quad amperes\ Total}$

Example

210.19(A)(1),(b)

Ambient Temperature – 35°C.-95°F.

Note: More than 10 feet – (3.0m) of the conductor, or more than 10% of the conductor length is within this ambient temperature (310.15(A)(2), Exception).

45 amperes (maximum load)

Table 310.15(B)(2),(a)

Temperature correction factor –   0.91 for 60°C Insulation   *(Table 310.15(B)(2),(a))*
0.94 for 75° C Insulation

$$\frac{45\,amperes}{0.91} = 49.45 \; amperes \; (49 \; amperes)$$

$$\frac{45\,amperes}{0.94} = 49.87 \; amperes \; (50 \; amperes)(220.5(B))$$

According to the general requirement of **210.19(A)(1),** the branch–circuit conductors must be sized to carry the larger of **210.19(A)(1),(a) or (b)(50 amperes or 53 amperes).** In this example, the larger ampere calculation is from **210.19(A)(1),(a) – 53** amperes

From **Table 310.15(B)(16),** a No. 6AWG (13.30 mm^2) copper conductor, with 60°C insulation (55 amperes), may be used **(110.14(C)(1),(a)).**

Branch Circuit Overcurrent Device

210.20(A)

The branch–circuit overcurrent device must be sized at 125% of the continuous load, plus any noncontinuous load.

In our example, this calculation resulted in a continuous and noncontinuous load of 53 amperes (30 amperes x 1.25, plus 15 amperes = 53 amperes).

The next higher standard rating of an overcurrent device is 60 amperes from 240.6(A). The conductor size from 210.19(A)(1),(a), based on the continuous and noncontinuous load, is 6 AWG copper-55 amperes – 60°C- Table 310.15(B)(16).

Section 240.4(B) permits a conductor with an ampacity that does not correspond to a standard overcurrent device rating (240.6(A)) to be protected with the next standard size overcurrent device. However, this permission does not apply where the rating exceeds 800 amperes, and where the conductors are part of a branch-circuit that supplies more than one receptacle for cord-and-plug connected portable loads.

In summary, this circuit consists of a 60 ampere fuse or circuit breaker, with a No. 6 AWG copper conductor. This complies with 210.19(A)(1),(a), 210.20(A), Table 310.15(B)(16), 240.4(B) and 110.14(C)(1),(a).

An Exception of 210.20(A) states the where an assembly with overcurrent devices for branch circuits is listed to operate at 100 percent of its rating, the ampere rating of the overcurrent device is permitted to be not less than the continuous load, plus any noncontinuous load.

Feeders

215.2(A)(1)

Once again, the feeder conductors supplying continuous loads or combinations of continuous and noncontinuous loads must have an ampacity of 125 percent of the continuous load, plus any noncontinuous load (215.2(A)(1),(a)).

If ambient temperature correction factors or adjustment factors where more than 3 current-carrying conductors are installed together (proximity effects), without maintaining spacing for lengths of over 24 inches (600 mm), and this causes the feeder conductor ampacity to exceed the ampacity calculation based on 125 percent of the continuous load, plus, any noncontinuous load, then this ampere rating must be used (215.2(A)(1),(b)).

Example

Feeder

Continuous load – 200 amperes
Noncontinuous load – 50 amperes

$$
\begin{array}{rl}
200 & amperes \\
\times 1.25 & (215.2\ (A)(1)(a) \\
\hline
250 & amperes
\end{array}
$$

$$
\begin{array}{rl}
250\ amperes & (Continuous\ Load) \\
50\ amperes & (Noncontinuous\ Load) \\
\hline
300\ amperes & Total
\end{array}
$$

Ambient temperature– 40^{0}C.– 104^{0}F

Note: More than 10 feet (3.0m) of the conductors, or more than 10% of the circuit length are within this ambient temperature (310.15(A)(2), Exception).

$$200 \; amperes$$
$$\underline{+50 \; amperes}$$
$$250 \; amperes \; Total$$

Table 310.15(B)(2)(a) ambient temperature correction factor 40°C. (104°F) ambient temperature 75°C. conductor temperature rating-Correction Factor – .88 (Table 310.15(B)(2)(a)).

$$\frac{250 \, amperes}{.88} = 284 \, amperes$$

The calculation for the continuous and noncontinuous load (300 amperes), exceeds the calculation based on the ambient temperature (40°C.), (284 amperes). So the feeder conductor ampacity must be based on the higher ampere value.

A 350 kcmil (177mm^2) copper conductor, which has a normal ampacity of 310 amperes at 75°C., is acceptable.

Feeder Overcurrent Protection

Where the feeder supplies continuous loads, or any combination of continuous and noncontinuous loads, the rating of the overcurrent device is based on 125 percent of the continuous load, plus the noncontinuous load (215.3).

In this example, the ampere rating will be the following.

$$200 \; amperes$$
$$\underline{\times \quad 1.25}$$
$$250 \, amperes$$
$$\underline{+ \quad 50 \, amperes}$$
$$300 \, amperes$$

Therefore, the feeder overcurrent device rating is 300 amperes with 350 kcmil (177mm^2) copper conductors.

Section 215.3 has an Exception which states that where the assembly, including the overcurrent devices protecting the feeder(s), is listed for operation at 100 percent of its rating, the ampere rating of the overcurrent device may be based on the sum of the continuous load, plus the noncontinuous load.

*Coordination (Selective)-

The definition of this term identifies a requirement to localize a fault (short-circuit or ground-fault), so as to restrict outages to a specific circuit or equipment. This is certainly desirable, not only to maintain a specific function, but, more importantly, to protect people. Selective coordination of protective devices is a requirement for Emergency Systems (700.32), Legally Required Standby Systems (701.27), and Critical Operations Power Systems, where overcurrent devices must be coordinated, but only for the period where the duration of the fault exceeds 0.1 seconds (0.167 cycles) (708.54). Also, review Section 517.31(G) for coordination of overcurrent devices for the Essential Electrical System of a Health Care Facility, and 620.62 for selective coordination of Elevator Feeders. For Critical Operations Data Systems and Multibuilding Campus-Style Complexes (Fire Pumps), Sections 645.27 and 695.3(C)(3) specify the requirements for selective coordination of overcurrent devices.

There is an important Informational Note in 230.95(C), Note 3 that states that where the Ground-Fault-Protection is provided for the service and there is another supply system through a transfer device (switch), other protection means or sensing devices may be necessary to assure proper ground-fault sensing by the GFP.

For Emergency Systems (700.6(D) or Legally Required Standby Systems (701.6(D), there will be a sensor for the ground-fault signal devices located at, or ahead of, the main system disconnecting means for the emergency system or the legally required standby system. This downstream sensing system must be selectively coordinated with the service GFP.

In addition, Section 517.17(C) applies to the selectivity of the required Ground-Fault Protection of equipment for the service and downstream feeder disconnecting means in a Health Care Facility. This selectivity is to assure that a ground-fault on the load side of the feeder disconnect will be isolated on this system without causing the service GFP to operate. This provision also applies to Critical Operations Power Systems (708.52(D), where the ground-fault protection must be fully selective.

Selective Coordination

Manufacturers of overcurrent devices provide valuable information to simplify a selective coordination study. One fuse manufacturer, Bussmann (Eaton) publishes Selectivity Ratios, based on line-side and load-size fuses, and these ratios assure selective coordination and they cover all circuit conditions from overloads to short-circuit conditions.

For example, a Fusetron RK 5 fuse used in conjunction with a Limitron. RK 1 fuse, where the fuse ampere ratings have a ratio of 1.5:1.

The ratios expressed in the Bussmann (Eaton) Fuse Selectivity Ratio guide are for all levels of overcurrent up to the fuse interrupting rating, or 200,000 amperes, whichever is lower.

This type of information certainly simplifies the selective coordination study.

Circuit breaker manufactures provide coordination tables to simplify this process, as well.

There are computer programs available that serve as an invaluable aid in selectively coordinating a series of overcurrent devices, based on the time-current curves published by manufacturers.

Before the analysis begins, the available fault-currents must be calculated at the source and downstream from the source. This, in itself, may require a significant effort, but it must be done.

Where ground-fault protection is required, the selective coordination study must assure that the ground-fault relays are properly set so that the ground-fault protection at the service is not affected when the ground-fault occurs on a downstream feeder (517.17(C), (708.52(D))).

*Demand Factor-

This is a concept where the total calculated load of several pieces of equipment may be reduced due to load diversity. For example, the demand factor for household electric ranges, wall-mounted ovens, counter-mounted cooking units, and other household cooking appliances with ratings over 1¾ kilowatts from Table 220.55. Also, Table 220.56, which covers demand factors for kitchen equipment in other than dwelling units, and demand factors for household electric clothes dryers in Table 220.54, as well as demand factors for non-dwelling receptacle loads from Table 220.44.

For example, in a non-dwelling, receptacle outlets are calculated at 180 volt-amperes each (220.14(I)).When calculating the total load for a group of receptacle outlets, let's say 500, a significant reduction in demand is permitted.

500 receptacle outlets (duplex) × 180va = 90,000 volt-amperes

Table 220.44

First 10,000 volt-amperes at 100% demand, and the remainder over 10,000 volt-amperes at 50% demand

$$\begin{array}{r} 10{,}000 \text{ VA} \\ + \underline{40{,}000 \text{ VA}} \text{ (50\% of the remaining 80,000 VA)} \\ 50{,}000 \text{ VA} \end{array}$$

So, the total load for the 500 receptacle outlets is 50,000VA. If supplied from a 240/120 volt, single-phase system, this load would be 25,000VA for each ungrounded conductor, or 208 amperes (25,000VA/120 volts).

If supplied from a 3-phase, 208/120 volt system, this load would be 16,667VA for each ungrounded conductor, or 139 amperes, per phase (16,667VA/120 volts, or 50,000VA/ 360.256 (208V × 1.732).

*Dusttight-

See Table 110.28, UL 1604-Dusttight is defined as 'enclosures constructed so that dust will not enter under specified test conductions' (ANSI/NEMA 250-2014).

A dusttight enclosure is typically NEMA - 3, 3S, 3SX, 4, 4X, 5, 6, 6P, 12, 12K, and 13. These enclosures may be suitable in Class II, Division 2 Locations (502.10 (B)(4)). Where enclosures are used in Class II, Division 1 Locations, they are classified as Dust-Ignition Proof, which are designed to preclude the entry of dust.

Explosionproof enclosures are not dusttight, because as the enclosure breathes due to atmospheric pressure differences, gases or vapors will flow through enclosure cover flanges, or other threaded joints, and dusts will also enter these enclosures. It is for this reason that explosionproof enclosures are not acceptable in Class II dust environments, underline they are identified for Class II, as well (502.5).

'Dust-Ignition Proof' is defined in Article 100. Enclosures identified for this purpose are classified as NEMA 9.

*Effective Ground-Fault Current Path-

This term defines the ground-fault conducting path, from the point where the ground-fault occurs, to the source of the electrical supply (transformer, generator, etc.).

This effective ground-fault current path must have 3 components. They are:

1. The path must be permanent and continuous.
2. The path must have ample capacity to safely carry the ground-fault current likely to be imposed on it. (This provision shows the importance of understanding the short-time ampere ratings of conductors, as the ground-fault current may be quite high, but typically of a short duration, until the overcurrent device operates to clear the ground-fault. The amount of ground-fault current that is expected to flow through this conducting path, and the duration of this ground-fault, or the time that it takes for the overcurrent device to clear this fault, must be known in order to satisfy this requirement.

As an example, the calculated short-circuit current on a 100 ampere circuit is 13,950 amperes. This circuit is protected with a 3-phase molded-case circuit breaker and the circuit conductors are 3 AWG THW, (26.67mm^2) copper. This circuit breaker has a fault clearing time of 0.025 seconds (IEEE 1584- Table 1).

The conductor insulation withstand rating of these circuit conductors is calculated as follows: one ampere for every 42.25 circular mils of conductor cross–sectional area for 5 seconds.

3 AWG – 52,620 circular mils (Table 8-Chapter 9)
I^{2}T (ampere-squared seconds) = one ampere for every 42.25 circular mils for 5 seconds.

$$\frac{52,620\, circular\ \ mils}{42.25} = 1245\, amperes$$

1245 amperes × 1245 amperes × 5 seconds = 7,750,125 ampere – squared seconds

$$\frac{7,750,125}{0.025\, seconds\ \ (clearing\ time\ of\ molded\ -\ case\, CB)} = 310,005,000$$

$$\sqrt{310,005,000} = 17,607\, amperes$$

The insulation on the 3 AWG –copper conductor will safely withstand 17,607 amperes for 0.025 seconds (1.5 cycles).

And, with the available short-circuit current at 13,950 amperes, the insulation on these circuit conductors will not be damaged, in compliance with 110.10.

If the equipment grounding conductor, (the effective ground-fault current path-250.4(A)(5)) is selected from Table 250.122, this conductor will be a minimum size 8 AWG-copper (8.37mm^2). In order to comply with 250.4(A)(5), this conductor must be 'capable of safely carrying the maximum ground-fault current likely to be imposed on it from any point on the wiring system where a ground-fault may occur to the electrical supply source'. This provision would apply for a system that is solidly grounded.

This leads to another consideration, that is, should this conductor be insulated or bare?

For an alternating current circuit, all of the conductors of the same circuit, including the equipment grounding conductor, must be within the same raceway, cable, cable tray, cord, trench, etc., in order to maximize the capacitive coupling effects between conductors and limit inductive heating in order to avoid increases in circuit impedance (300.3(B)), (300.5(I)).

Ground-faults are the most common type of fault. In some instances, studies have shown that 90% of faults in distribution systems are ground-faults. And 90% of these ground-faults are arcing type faults, where the ungrounded conductor is not in direct contact with the equipment grounding system. And, typically, an intermittent arc develops from the ungrounded conductor to the equipment grounding system. To make matters worse, there is a voltage-drop in the arcing fault which will reduce the ground-fault current returning to the source. Certainly, this condition will have an effect on the operation of the circuit overcurrent device, and it must be considered.

A bolted ground-fault, that is, the equivalent of the ungrounded conductor physically bolted to the equipment grounding conductor will produce a current flow of 100% of the available fault-current near the source (transformer, generator, etc.), to 50% of the available fault-current downstream from the source.

An arcing ground-fault will produce a fault current of a <u>minimum</u> of 38% of the available fault-current, but using 50% would be better.

If we use 50% of the available short-circuit current in our example (50% of 13,950 amperes), this would equal 6975 amperes.

Deciding on whether to use an insulated copper equipment grounding conductor or a bare copper conductor is an important consideration. Equipment grounding conductors may be solid or stranded, insulated, covered, or bare (250.118(1)). 6975 amperes of ground-fault current, even for a brief period, in this case for 0.025 seconds, will produce a great deal of thermal stress on the other conductors within the same raceway, cable, etc. An insulated equipment grounding conductor will provide better protection against thermal stress than one that is bare.

But, will 6975 amperes for 0.025 seconds damage the insulation on this conductor?

$$8AWG - copper - \frac{16,510\,circular\,mils}{42.25} = 391\,amperes$$

391 amperes × 391 amperes × 5 seconds = 764,405 ampere-squared seconds

$$\frac{764,405}{0.025\,seconds} = 30,576,200$$

$$\sqrt{30,576,200} = 5530\,amperes$$

The insulation withstand rating for the 8 AWG copper conductor for 0.025 seconds is 5530 amperes, and the available ground-fault-current has been determined to be 6975 amperes. So, based on these conditions, the insulation would be extensively damaged, and this constitutes a violation of 110.10.

If we increase the equipment grounding conductor size to 6 AWG-copper, we would increase the conductor insulation withstand rating as follows:

$$\frac{26,240\,circular\,mils}{42.25} = 621\,amperes$$

621 amperes × 621 amperes × 5 seconds = 1,928,205 ampere-squared seconds

$$\frac{1,928,205}{0.025\,seconds} = 77,128,200$$

$$\sqrt{77,128,200} = 8782\,amperes$$

The insulation withstand rating for the 6 AWG copper conductor for 0.025 seconds exceeds 50% of the available short-circuit current of 13,950 amperes (6975 amperes). So, no damage to the insulation on the equipment grounding conductor during the clearing time of the molded-case circuit breaker and no violation of 110.10.

If the effective ground-fault current path is based on the <u>fusing</u> or <u>melting</u> current of the 8 AWG copper conductor, as opposed to the insulation withstand rating of this conductor for 0.025 seconds, the calculation is as follows:

$$8\,AWG - copper\frac{16,510}{16.19} = 1020\,amperes\,(5\,seconds)$$

I^2T = one ampere for every 16.19 circular mils for 5 seconds.
1020 amperes × 1020 amperes × 5 seconds = 5,202,000 ampere–squared seconds

$$\frac{5,202,000}{0.025\,seconds} = 208,080,000$$

$$\sqrt{208,080,000} = 14,425\,amperes$$

14,425 amperes is above 50% of the available short-circuit current of 13,950 amperes, or 6975 amperes.

Considering the fusing or melting current, as opposed to the insulation withstand rating of the equipment grounding conductor, the 8 AWG copper conductor from Table 250.122, would be acceptable.

Another possible solution would be to use a current-limiting circuit breaker instead of the standard device. Current-limiting circuit breakers will clear this ground fault within ½ cycle (0.008 seconds), when interrupting currents within their interrupting range. So, instead of a 1½ cycle (0.025 seconds) clearing time for a standard molded-case circuit breaker, we would have reduced the fault-clearing time to ½ cycle. And, this would increase the insulation withstand rating for the 8 AWG copper conductor (8.37 mm^2) to 9,775 amperes. And, with an available ground-fault current of 6975 amperes, this conductor will provide an effective ground-fault current path in compliance with 250.4(A)(5).

If the effective ground-fault current path is in the form of a metal conduit, either in parallel with an internal equipment grounding conductor (250.134(B)), (300.3(B), or as the sole effective ground-fault current path, the ability of this conductor to satisfy the provisions of 250.4(A)(5) must be determined.

Information on this topic is available, such as, 'Modeling and Testing of Steel, Intermediate Metal Conduit, and Rigid (GRC) Conduit, Part I, May 1994, Copyright 1994, Georgia Tech Research Institute'.

This is a good example of using the conductor short-time current rating charts in this book to determine the validity of the effective ground-fault current path.

3. The effective ground-fault current path must be of sufficiently low impedance to limit the voltage-to-ground and facilitate the operation of the overcurrent device (on a solidly-grounded system). During a ground-fault, as the equipment grounding conductor is carrying the ground-fault current back to the source (transformer, generator, etc.), the voltage-drop associated with this current flow produces a voltage-rise, above earth potential (0 volts), on anything connected to the equipment grounding conductor. This may produce an excessive voltage-rise on equipment. Possibly a dangerous touch-potential may appear on equipment frames. So, the equipment grounding conductor must be properly sized to limit the voltage-drop and resultant voltage-rise.

Where a change occurs in the size of the ungrounded conductors to compensate for voltage-drop, a proportional increase must be made in the size of the equipment grounding conductor (250.122(B)), unless the Authority Having Jurisdiction permits a qualified person to determine the size of the equipment grounding conductor .

(Check the UL guide on FHIT cable systems (e.g., cables that are installed on the life/safety and critical branches of an emergency system), where the equipment grounding conductors are installed in raceways and the equipment grounding conductor and raceway is a part of the electrical circuit protection system).

Example

Branch Circuit Rating - 40 amperes
Circuit Voltage - 240 Volts
Single-Phase Circuit-Load 32 amperes (Continuous)
Normal Ungrounded Conductor Size - 8 AWG, Copper
Minimum Equipment Grounding Conductor Size - 10 AWG Copper
(Table 250.122)
Circuit Length (From source to load) - 170 feet - (51.80 meters)

According to Section 210.19(A), Informational Note 4, conductors for branch circuits that are sized to prevent a voltage-drop exceeding 3% at the farthest outlet of power, heating, and lighting loads, or combinations of such loads, and where the maximum total voltage-drop on a combination feeder and branch circuit is limited to 5% of the applied voltage, will provide for reasonable efficiency of operation.

For a feeder, this Informational Note expresses a voltage-drop of 3% of the applied voltage as the recommended limit for reasonable efficiency of operation (215.2(A)(1)(b), Informational Note 2. And, the voltage-drop for a combination feeder and branch-circuit should be limited to 5%.

Remember, as Informational Notes, this information is not mandatory (90.5(C)). So a lower, or even higher level of voltage-drop may be acceptable, or even necessary, depending on the type of equipment supplied.

$$Voltage - drop = \frac{2K \times L \times I(Single - Phase)}{\text{circular mil area of conductor}}$$

$$Voltage - drop = \frac{1.732 \times L \times I \ (Three - Phase)}{\text{circular mil area of conductor}}$$

Where
K = 12.9 - copper wire
K = 21.2 - aluminum wire
K= equals the conductor resistance in ohms per 1000 feet, multiplied by the conductor circular mil area, divided by 1000 (Table 8- Chapter 9 – Informational Note).

Example – Table 8 – Chapter 9

1 AWG copper (uncoated) – 83,690 cm
DC resistance at 75°C. – 1000 feet - .154 ohms

$$83,690\,cm$$
$$\underline{x.154}$$
$$12,888.26$$

$$\frac{12,888.26}{1000} = 12.88826\,or\,12.9\,ohms$$

K = ohms per cm foot, or ohms per mil foot
Cmil = wire diameter in inches or decimal inches $\times 1000^2$

Example

Wire diameter – ¾ (.75) inch
.75 × 1,000,000 (1000^2) = 750,000cm (750kcmil)

L = One-way length in feet of conductor, from the source to the load.
I = amperes of the load

$$Voltage - drop = \frac{25.8 \times 170\,feet \times 32\,amperes}{16,510\,circular\,mils\,(No.\,8\,AWG)\,(Table\,8\,-\,Chapter\,9)} = 8.5\,volts$$

$$3\%\,of\,240\,volts = 7.2\,volts$$

So, we increase our branch circuit conductor size from 8 AWG copper to 6 AWG copper to reduce voltage-drop.

$$Voltage - drop = \frac{25.8 \times 170\,feet \times 32\,amperes}{26,240\,circular\,mils} = 5.35\,volts$$

$$\frac{6\,AWG - 26,240\,circular\,mils}{8\,AWG - 16,510\,circular\,mils} = 1.59$$

Minimum Equipment Grounding Conductor Size for 40 ampere circuit -10 AWG copper (Table 250.122) 10 AWG - 10,380 circular mils × 1.59 = 16,504 circular mils-(8 AWG = 16,510 cm)

The Equipment Grounding Conductor for this circuit is 8 AWG copper. This satisfies the proportional increase in the equipment grounding conductor size to comply with Section 250.122(B). This method has been used for many years, unless the AHJ permits a qualified person to determine the EGC size.

*Explosionproof Equipment-

This type of equipment is designed to breathe through its cover and other threaded entry joints. Ignitable concentrations of gases or vapors are often likely to form inside of this equipment. If these gases or vapors are ignited within this enclosure, they would be sufficiently cooled, as the gases expand from the point of ignition, and by the enclosure itself, before they are released into the external atmosphere. So the enclosure joints are 'flame-arresting.'

In addition, the enclosure is designed to withstand at least four times the maximum explosion pressure created within it, without rupturing, or without permanent distortion.

One further note, this enclosure is designed to operate so that its external temperature will not become a source of ignition.

Explosionproof enclosures are not dusttight or dust-ignition proof. Where these enclosures have flanged covers with flat cover joints, there is a purposeful gap in the cover joint of .0015" (0.04mm). Even enclosures that have threaded entries are not, necessarily, dust-ignition proof. Explosionproof enclosures are designed to breathe, and this process occurs due to differing ambient temperatures from inside to outside of the enclosure. And, the atmospheric pressure differences that are created by these lower and higher temperatures. The flat cover joints or the threaded cover joints are designed to be flame-arresting. This means that if ignition of a flammable gas or vapor occurs within the enclosure, the heated ignition gases are cooled by the mass of the enclosure as they pass through the flanged cover joint, or the threaded cover joint, before being released into the outside atmosphere. In fact, the five fully engaged, threaded joint connection for threaded entries into explosionproof equipment, as referenced in Section 500.8(E)(1), are based on the concept of providing a flame-arresting path. Explosionproof enclosures with threaded covers have at least 4½ fully-engaged threads for the same reason, (500.8(E)(1), Exception).

*Exposed (as applied to wiring methods)-

On, or attached to the surface, or installed behind panels (such as removable ceiling panels) designed to allow access. So, the concept of being exposed, does not always mean 'normally visible' (300.22 and 300.23).

*Feeder-

The conductors extending from the service equipment, or from the source of a separately-derived system, or another power supply system, and the final branch-circuit overcurrent device. Articles 215, 220, and 225 address the application of 'feeders'.

Feeder conductors do not directly supply utilization equipment. Branch-circuit conductors supply this equipment, regardless of the size of the circuit.

Beyond the service equipment, it is relatively common for feeder conductors to supply transformers, and the feeder conductor size is based on the nameplate rating of the transformer(s), when only transformers are supplied (215.2(B)(1)).

Where the feeder conductors supply transformers and utilization equipment, their size is based on the nameplate rating of the transformer(s), plus 125% of the load of the utilization equipment, where both the transformer(s) and utilization equipment will operate simultaneously (215.2(B)(2)).

Feeders that supply 15 and 20 ampere receptacle branch circuits, where ground-fault circuit interrupter protection is required by 210.8 and 590.6(A) (temporary wiring installations), may provide the GFCI protection for the downstream branch circuits. The feeder GFCI must be in a readily accessible location (215.9).

However, where the feeder provides this protection for several branch circuits, a single ground-fault may interrupt all of the protected branch circuits.

*Field Evaluation Body-

An organization or part of an organization (e.g., UL) that performs field evaluations of electrical equipment.

Field evaluation has become common in recent years, as Authorities Having Jurisdiction will typically require an evaluation when equipment is modified, or when environmental conditions change, or when no listing or certification exists for equipment that has been installed, or is about to be installed (NFPA 790).

*Field Labeled as Applied to Evaluated Products-

Equipment or materials to which has been attached a label, symbol, or other identifying mark of a Field Evaluation Body, indicating the equipment or materials were evaluated and found to comply with requirements as described in an accompanying Field Evaluation Report.

*Fitting-

An accessory, such a locknut, bushing, or other part of a wiring system that is intended primarily to perform a mechanical rather than an electrical function. Locknuts, bushings, cable connectors, and conduit bodies are examples of fittings.

As we begin our focus on the topic of grounding systems and equipment, it is important to understand the reasoning behind the purpose of using proper methods to assure that the system and equipment provide the appropriate safety and efficiency of operation that we are striving to achieve.

The two main functions of grounding can be summarized as follows:

- **To provide a means to limit exposure to the higher voltages that are associated with lightning or other external power faults.**
- **To provide a means to stabilize system voltages and their relationship to ground (earth) during normal and abnormal conditions.**

*Ground-

The earth (or to some conducting body that serves in place of the earth).

*Ground-Fault-

This is an accidental connection between an ungrounded conductor and an equipment grounding conductor, which may be a wire, metal raceway, metal cable tray, metallic equipment frame, or the earth.

Based on extensive research, it is safe to conclude that 90% of faults occurring in electrical systems are ground-faults, and, further, 90% of ground-faults are arcing-type faults. In addition, there is a voltage-drop associated with this arcing fault, which would have an effect of reducing the amount of ground-fault current in this circuit. And, couple this problem with the impedance of the ground-fault return path, and it is easy to see why the circuit overcurrent device may not operate in a short enough period of time, or, at all, to prevent a hazard to people, and/or, to equipment. And because the ungrounded conductors come in contact with the equipment grounding system in many places throughout a typical distribution system, it is easy to see why a ground-fault is the most common type of fault. It may not be caused by damage to the insulation, but simply, through contaminants that have been absorbed through the insulation, such as dirt combined with moisture.

*Grounded-

This is a connection to ground or to a conducting body that connects to ground (earth).

*Grounded, Solidly-

This is a physical connection to ground (earth) through a connection without a resistance element or impedance device in series with this connection. (250.20(A),(B),(C)).

*Grounded Conductor-

This is a conductor that has been intentionally grounded. This conductor is, most often, referred to as the 'neutral' conductor. On single-phase, three-wire systems, three-phase, four-wire, Wye connected systems, and three-phase, 4-wire Delta systems, the system <u>neutral point</u> is solidly connected to ground (earth). The neutral point of these systems is the physical point, where voltage from the other points of the system would be equal. For example, a three-phase, Wye-connected transformer secondary where voltage from X1-X2-X3 to the X0 (neutral point) is the same. The X0 (neutral point) would be solidly grounded, and a neutral (grounded) conductor would originate from X0 (250.24, 250.26).

However, on a three-phase, Delta, corner-grounded system, the grounded conductor is not a neutral conductor.

So, we can say, that all 'neutral' conductors are grounded conductors, but not all grounded conductors are 'neutral conductors'.

Section 200.4 applies to the installation and identification of neutral (or grounded) conductors.

Sections 200.6 and 200.7 specify the requirements for identifying 'grounded' conductors, typically by the use of 'white' or 'gray' insulation, or by 'white' or 'gray' markings at terminations.

*Ground-Fault Circuit Interrupter-

This device has been available since 1965, and was first referenced in the 1971 NEC. It has a differential transformer, which surrounds the supply and return conductors, and monitors the current difference in these wires. If the current difference reaches a level of 4-6ma, or higher, the unit trips in about 1/40 second. The listing standard for this device is UL 943.

The required use of these devices has expanded over time. One of the original code references, which addresses the use of GFCI's in dwelling and nondwelling units, is Section 210.8.

NEC References for GFCI Requirements

210.8(A) – Dwelling units

210.8(A)(3)– All dwelling unit outdoor receptacle outlets-125-250 volt and 150 volts-to-ground, or less

210.8(B) – Bathrooms, in other than dwelling units

210.8(B) – Rooftops, in other than dwelling units

210.8(B) –Nondwelling laundry areas within 6 feet of the outside edge of a bathtub or shower stall

210.8(B) –Nondwelling kitchen that has a sink and a permanent means for preparing or cooking food

210.8(F) – All outdoor outlets from single-phase branch circuits up to 150 volts-to-ground and 50 amperes, or less.

215.9 – Feeders (supplying 15 and 20 ampere receptacles)

422.5 – Appliances (Move to Article 422-2020 NEC)

424.44–45–99 – Fixed electric space heating equipment

445.20 – Generators

513.12 – Aircraft hangers

517.20(A) – Health care facilities

525.23 – Carnivals, circuses, fairs, and similar events

550.13(B) – Mobile home – receptacle outlets

550.32(E) – Mobile home – outside receptacles

551.40(C) – Recreational vehicles-internal wiring protection

551.40(C) – Recreational vehicles-receptacle outlets

552.41(C) – Park trailers-receptacle outlets

555.19(B) – Marinas, Boatyards, Boathouses (other than shore power)

590.6 – Temporary wiring on construction sites

600.10(C)(2) – Electric signs-portable and mobile

620.85 – Receptacle outlets–elevators, escalators, moving walks, platform lifts, and stairway chairlifts

626.24(D) – Receptacle outlets-electrified parking space

647.7(A) – Sensitive electronic equipment

680.5, 680.6, 680.22(A)(4), 680.22(B)(4), 680.32, 680.62(E) – Pools

680.71 – Hydromassage bathtubs

680.27(B)(2) – Electrically operated pool covers

682.15 – Receptacle outlets-naturally and artificially made bodies of water

*Ground-Fault Current Path-

This is defined as a conductive path, from the point on the wiring system where the ground-fault develops, and then, through any conducting paths back to the source of the electrical supply. Hopefully, the lowest impedance path for this current to flow is through the equipment grounding conductor. However, other conducting paths may parallel the equipment grounding conductor. A portion of the ground-fault current may flow through these parallel paths. Also, when the ground-fault current reaches the service equipment, the current will divide, unevenly, on its way back to the source. Due to the fact that there will be a grounding electrode connected to the grounded conductor and the equipment grounding system within the service equipment, or ahead of the service equipment, and, another grounding electrode connected to the neutral or grounded point of the supply transformer, another parallel path is established through the earth between these two grounding electrodes. Some of the normal current flowing through the grounded conductor will return to the transformer through this earth path, as well as some of the ground-fault current. This is a conducting path, albeit, a high impedance path, as compared to the path through the service neutral or grounded conductor.

*Ground-Fault Protection of Equipment-

This protection is a critical part of a solidly grounded, 3-phase, 4-wire, Wye connected system which operates at over 150 volts-to-ground, and not in excess of 1000 volts phase-to-phase. The most common type of system for this protection is a three-phase, 4-wire, solidly grounded, Wye electrical system with a voltage of 480 volts, phase-to-phase, and 277 volts, phase-to-neutral, and having a service disconnect rated at 1000 amperes, or more. This is the type of distribution system where arcing ground-faults are the most destructive.

It is very important that the pick-up setting of the GFP is selectively coordinated with downstream overcurrent devices in order to prevent a nonorderly shutdown of an entire system. Sections 215.10 (Feeders), 230.95 (Services), 517.17 (Health Care Facilities) and 708.52(D) (Critical Operations Power Systems) address the requirements for this protection. Sections 517.17(B) and 708.52(B) require a second level of ground-fault protection on every feeder downstream of the protection at the service equipment. This downstream GFP will have a lower pickup setting than the main GFP to assure that the feeder device will open ground-faults on its load side, without affecting the operation of the service GFP device. This will limit an outage to a single feeder, so that the rest of the distribution system will remain operational.

*Grounding Conductor, Equipment-

This is a conductor, which may be in the form of a wire, metal raceway, metallic cable assembly, or metal cable tray. Section 250.118 identifies the various types of equipment grounding conductors. These conducting paths must be able to carry the maximum ground-fault current that may be imposed on them. If the equipment grounding conductor is in the form of a wire, Table 250.122 specifies its minimum size, in conjunction with Sections 250.4(A)(5) and 250.4(B)(4). Section 250.119 covers the identification of an insulated equipment grounding conductor. It is to have a continuous outer finish that is green, or green with one or more yellow stripes. However, where conductors are 4 AWG (21.15 mm^2) or larger, other means of identification are acceptable, such as the use of green tape at terminations and other places where the wire is accessible, except in fittings, such as conduit bodies, that contain no splices or unused hubs.

Sections 300.3(B) and 300.5(I) require that all of the conductors of the same circuit be installed together, and in close proximity to each other to maximize the effects of magnetic and capacitive coupling and reduce the overall circuit impedance. 690.43(C) requires the equipment grounding conductors for the photovoltaic array and support structures to be within the same raceway or cable or otherwise run with the PV array circuit conductors where these conductors leave the vicinity of the PV array

*Grounding Electrode-

This is the object that establishes a means of making a connection to the earth.

Section 250.52 identifies the various types of grounding electrodes that are used to make this earth connection.

It is very important that the grounding electrode, or grounding electrode system selected for a particular installation, be of a type that will assure a low resistance-to-ground. This concept is not only relative to the safety of people, but also, to assure the proper operation of equipment. This is especially true, due to the sensitivity of the sophisticated equipment in use today. And this may be more difficult now, as this equipment may be installed in an existing installation where a proper grounding system was not considered a priority.

In the past several years, I have had the privilege of working in many countries throughout Africa, as well as Haiti, and, even in several locations in North Korea. The installations were all health care facilities. In the vast majority of these installations, a proper grounding system was not available, and, in some instances, there was no grounding system at all.

As an example, I was called upon to check the grounding system at the oldest government hospital in Addis Ababa, Ethiopia. GE Medical Systems was providing a new CT Scanner for this hospital, and it has been my experience in other installations, that GE requires a grounding electrode system to have a maximum resistance-to-ground of less than 2 ohms. Depending on the type of soil, as well as other conditions, this low resistance-to-ground requirement can be difficult to achieve. I solved this problem at another hospital in Ethiopia through the use of a ground ring, supplemented by several driven ground rods. But, that was a new installation, and there were no area restrictions. This government hospital in Addis Ababa was in a confined area, so a ground ring was not an option. My preferred grounding electrode would have been a concrete-encased electrode. But this was an existing building, so there was no access to the rebar in the concrete footings (250.50, Exception).

Another option, that I had used at other installations, was a chemically-charged ground rod, which may have produced the desired less than 2 ohm resistance-to-ground. But, this type of rod (UL-467J) was not available in Ethiopia, and, to have one shipped from the U.S. would have taken at least six months, as the customs process in Ethiopia is, frustratingly, slow.

So, I decided to install a group of driven ground rods, 10 feet (3.048 meters) long × 5/8" (15.875 mm) in diameter. These were copper-clad steel rods, and they were properly spaced a minimum of 2 rod lengths, or 20 feet (6.096 meters), (see the Informational Note following Section 250.53(A)(3)). In order to achieve a resistance-to-ground of less than 2 ohms, I had to install 13 ground rods. An engineer from GE performed a ground-resistance test to verify this low resistance-to-ground.

This may sound unrealistic, but before the installation of a driven or buried electrode, it is important to perform a soil-resistivity test in order to determine the best location for the grounding electrode or system.

There are three components that affect the resistance-to-ground of a driven or buried grounding electrode.

1. The metallic mass of the electrode. This portion of the resistance-to-ground is negligible. For example, the electrical resistance of a typical ground rod is quite low. A typical 5/8 inch (15.875mm) ground rod has the equivalent current-carrying capacity of a 3/0 AWG ($85mm^2$) copper conductor.
2. The resistance of the metal/soil interface. This is also quite low, unless there is a nonconductive coating on the metal mass of the electrode.
3. The resistivity of the soil surrounding the electrode. This is the component that will determine the resistance-to-ground of the grounding electrode. And, soil resistivity varies from place to place, and in some instances, even in the same area.

It would certainly be beneficial, and yes, very often critical, to locate an area where the soil resistivity is the lowest of the surrounding area, as this is the best location for the electrode. The soil resistivity is expressed in ohm-centimeters. An ohm-centimeter is the electrical resistance across the faces of a cubic centimeter of soil. Where the soil resistivity is high, or unknown, soil enhancements are available. Bentonite is a natural clay and has been used for this purpose. This material contains aluminum, iron, magnesium, and sodium. When this material comes in contact with water, it expands. Buried conductors may be surrounded with this material. Just like the typical soil surrounding a driven ground rod, the material within 6 inches (152.40mm) of the rod or wire will have a dramatic effect on the total resistance-to-ground of the rod or conductor.

Erico 'GEM' (Ground-Enhancement Material) is even a better material, as its resistivity is even lower than Bentonite (Erico.com). And, it adheres to the surface of the rod or conductor much better than Bentonite. In addition, this material is noncorrosive.

Section 250.52(A) lists the acceptable grounding electrodes. They are:

1. A metal underground water pipe: In order to qualify as a suitable grounding electrode there must be at least 10 feet (3.048 meters) of metallic underground water pipe in direct contact with the earth. Since the 1978 NEC, underground metal water pipes must be supplemented by at least one additional grounding electrode. This is, very often, a ground rod. The reasoning here is that even though the metal underground water pipe may be extensive, and, as such, the resistance-to-ground of this pipe may be quite low, if repairs to this pipe become necessary, some metallic portions of the pipe may be replaced with nonmetallic pipe. And, if the supplemental electrode is a ground rod, the resistance-to-ground of the single rod, pipe, or plate must not exceed 25 ohms. If so, these grounding electrodes must be supplemented by an additional grounding electrode (250.53(A)(2), Exception).
2. The metal frame of a building or structure which has been effectively grounded. This includes at least one metal member of the building that is in direct contact with the earth, vertically, for 10 feet (3.048 meters), or more. This structural metal member may, or may not, be encased in concrete. Or, the tie-down bolts which secure a structural metal member (steel column) to a concrete footing, with the tie-down bolts connected to the reinforcing rods in the footing.
3. A concrete-encased electrode: This grounding electrode is my personal favorite, as there is typically a very large amount of reinforcing rod,

encased in concrete, which is in direct contact with the earth. As long as the concrete is not separated from the earth by a vapor barrier, or another type of nonconducting coating, this grounding electrode has a very low resistance-to-ground. A section of rebar may extend from the footing or foundation (above grade level) for the attachment of the grounding electrode conductor. The connector must be listed and identified for a copper to steel connection. Or, it may be exothermically welded.

4. Ground Ring: This is the grounding electrode that I used for the new radiology building at Soddo Christian Hospital in Ethiopia. A new digital X-ray machine and a new CT Scanner were to be installed, as well as 2-UPS units, and possibly, an MRI sometime later. Once again, GE required a grounding electrode system with a resistance-to-ground of less than 2 ohms. The concrete footings were already poured, so I decided to use a ground ring. The ring consisted of a No. 2 AWG (33.63 mm^2) solid copper wire, buried to a depth of 3 feet (1 meter). The ground ring was supplemented by 8-5/8″ × 10' ground rods. The length of the ground ring conductor was 300 feet (91.44 meters). A 3-phase, 380/220 Volt, 165 kVA generator was installed at a later date. And, the generator neutral point was bonded to this ground ring. So everything in that building is at, virtually, the same potential.

5. Rod and Pipe Electrodes: Ground rods are very common, but also, they may be very ineffective. They are tested in accordance with UL 467. A copper-clad steel rod has an <u>average</u> service life of 30 years. This service life is dependent upon soil conditions, in that, if the soil has a low-resistivity, let's say, less than 10,000 ohm-centimeters, it is considered highly corrosive. From 10,000-30,000 ohm-centimeters, the soil is considered mildly corrosive. And above 30,000 ohm-centimeters, it is considered noncorrosive.

   The ground rod is driven to a depth of at least 8 feet (2.44 m), except where rock bottom is encountered, it may be driven at an angle up to 45° from the vertical, or buried horizontally in a trench at least 30″ (750mm) deep. And, typically 5/8″ (15.87mm) in diameter, although 'listed' ground rods may be ½ ″ (12.7 mm) in diameter.

   A ¾ ″ trade size (metric designator 21) galvanized pipe may be used, with the same minimum conditions as the rod. However, the service life of this grounding electrode may be limited, as compared to a copper-clad steel rod.

6. Other Listed Electrodes are also permitted, such as the chemically-charged rod that I spoke of earlier (UL 467J). These are hollow core rods, and they are filled with a low resistivity material, which may be

magnesium sulfate. The rod may have an outside diameter of 2⅛″ (53.975 mm), and they are normally 10 feet (3.048 meters) in length. Although they can be made longer, and I have used them in an 'L' shape configuration. This low resistivity material reacts with the normal moisture in the air and dissolves and will slowly migrate into the soil through a series of holes in the shaft of the rod, thereby adding minerals to the soil and lowering the soil resistivity. The performance of this rod will increase over time.

These rods may have a service life of up to 50 years. Where the soil resistivity has not been tested, I would strongly recommend that these electrodes be used.

7. Plate Electrodes: Plate electrodes, if of coated iron or galvanized steel are required to be at least ¼″ (6.4 mm) in thickness. If of nonferrous material, they may be 0.06″ (1.5 mm) in thickness. They must expose at least 2 square feet (0.186 m²) to exterior soil. These plate electrodes, while not commonly used, are typically buried on end, as opposed to lying flat, in order to reduce the size of the excavation.

250.52(B) addresses the materials and systems not permitted to be used as grounding electrodes.

1. A metal underground gas piping system. This is not due to an explosion hazard. It is because of the impressed-current cathodic protection system that the gas supplier may be using to protect the gas pipe from galvanic corrosion. Connecting a grounding electrode conductor to the gas pipe could negate this protection due to the low level of AC current flowing through the grounding electrode conductor, and into the gas pipe, and then through the earth on its path back to the supply transformer.

   Also, the gas pipe will have insulating joints. So, a long uninterrupted metal pipe will not be available.

2. Aluminum electrodes: Due to corrosion and oxidation, these electrodes would have a limited service life.

3. The bonding grid of a swimming pool (680.26) may not be used as a grounding electrode due to the hazard of elevated potential differences in the pool area (250.52(B)(3)).

Section 250.53(A)(2) states that a single rod, pipe, or plate is to be supplemented by an additional grounding electrode, and the supplemental electrode may be any of the types referenced in Section 250.52(A)(2) through (A)(8). There is an exception here that waives this requirement if the single, rod, pipe, or plate has a resistance-to-ground (earth) of 25 ohms, or less. I have always maintained that this 25 ohm value is too high to provide proper protection for people, systems, and equipment. Especially, if there was a problem with lightning. If lightning were to strike at the electrical distribution system, or, more likely at an outside transformer supplying this distribution system, and the lightning current, of say, 20,000 amperes was carried into this ground rod with a 25 ohm-to-ground resistance, the instantaneous voltage rise, above earth potential, would be 500,000 volts. While this would be an instantaneous event (lightning currents reach a peak in about 2-10 microseconds), this, certainly, could be hazardous to people, and damaging to equipment.

Section 250.53(D)(2) requires that a metal underground water pipe be supplemented by an additional electrode of a type specified in Section 250.52 (A)(2) through (A)(8).

It is extremely important that where multiple grounding electrodes are installed that they be bonded together to form one common grounding electrode system (250.50), (250.58). This includes the grounding electrodes that are installed for the strike termination devices of a lightning protection system (250.60),(250.106). The bonding of all separate grounding electrodes will serve to limit potential differences between them, and the systems and equipment which are connected to ground through these grounding electrodes.

Another benefit associated with the bonding of the grounding electrodes of different systems is that the overall resistance-to-ground of all of the connected systems may be markedly reduced, as these systems will now be in parallel. This will provide added protection for all of these systems.

*Grounding Electrode Conductor-

The grounded conductor and the equipment grounding conductor are connected to the grounding electrode (system) by the grounding electrode conductor. Section 250.62 states that this conductor may be copper, aluminum, or copper-clad aluminum. And the conductor may be solid or stranded, insulated, covered, or bare.

Generally, this conductor must be installed in one continuous length. However, exothermically welded connections or compression (crimped) type connections are acceptable, as these connections are irreversible, which means that they cannot be taken apart with tools. Busbar connections are also acceptable (250.64(C)).

Where multiple service disconnects are installed as a group in separate enclosures (230.72(A)), taps from each disconnect to the main grounding electrode conductor are permitted. This would be common where these separate enclosures are connected to an auxiliary gutter, with the splices inside this gutter (250.64(D)).

Enclosures for grounding electrode conductors should be nonmetallic. However, steel enclosures are acceptable for physical protection if this enclosure is bonded on both ends to the internal grounding electrode conductor. This may be accomplished through the use of bonding bushings on each end of the steel enclosure. The size of the bonding jumper would be the same size as the internal grounding electrode conductor. The reason for this requirement is due to the magnetic properties of the steel enclosure. If a significant current was flowing through the grounding electrode conductor, the magnetic field surrounding this conductor would induce a current in the steel enclosure, which would be in a direction opposite the current flow through the grounding electrode conductor. This condition would produce an 'inductive choke' on the grounding electrode conductor, which would significantly increase the impedance of the earth connection (250.64(E)). Possible damage to the grounding electrode conductor, due to arcing within, or at the end of the metal pipe, may occur.

Where the steel enclosure is bonded to the internal conductor on both ends, the effects of this 'inductive choke' are greatly reduced.

If nonferrous enclosures are used for the physical protection of the grounding electrode conductor, bonding the conductor on each end of the enclosure is not necessary, due to the nonmagnetic properties of this material (250.64(E)(1)).

The sizes of AC Grounding Electrode Conductors are listed in Table 250.66. for connections to metal underground water pipes and metal in-ground support structures. This size is based on the size of the ungrounded service-entrance conductors, or the size of the ungrounded feeder conductors that extend from the source of a separately-derived system. From data that I studied 40 years ago, I have maintained that the wire sizes expressed in Table 250.66 are based on lengths up to 100 feet (30.48 meters), and a specific voltage-drop that is associated with this length. Because this voltage-drop is directly proportional to the voltage-rise on connected systems and equipment during fault conditions, the grounding electrode conductor should be increased in size for lengths of over 100 feet (30.48 meters). But, due to the inductive reactance associated with the increased frequency of lightning, the length of the grounding electrode conductor should not be any longer than necessary to make the connection to the grounding electrode (250.4(A)(1)), Informational Note No. 1).

A No. 6 AWG copper (13.30mm) or No. 4 AWG aluminum (21.66 mm) conductor may be used as the sole connection to a ground rod, pipe, or plate electrode (250.66(A)). However, the use of an aluminum conductor is restricted due to the effects of corrosion. Aluminum conductors or copper-clad aluminum conductors are not permitted within 18″ of the earth (250.64(A)).

For concrete-encased electrodes, the connection to the reinforcing steel within a concrete footing may be No. 4 AWG copper (21.66 mm^2) (250.66 (B)).

For connection to a ground ring, (250.52(A)(4)), the conductor may not be smaller than 2 AWG bare copper, at least 20 feet (6.0m) long, and at least 30 inches (750 mm) below the surface of the earth (250.53(F)).

Where multiple grounding electrodes are connected, e.g., a ground ring bonded to a concrete–encased electrode and to a ground rod. The grounding electrode conductor size is based on the largest conductor required among all of the electrodes connected to it.

In this case, a minimum size of 2 AWG copper for the ground ring (250.64(F)).

The size of the grounding electrode conductor for direct current systems is specified in 250.166(A),(B),(C),(D), and (E)

250.66– This Section and Table identifies the sizes of grounding electrode conductors for alternating current systems (250.166 identifies the sizes of grounding electrode conductors for direct-current systems). For Rod, Pipe, or Plate Electrodes, (250.66(A)), the grounding electrode conductor is not required to be larger than 6 AWG copper, if this conductor is the sole connection to the rod, pipe, or plate. A 5/8″ (0.15875 meters) diameter copper-clad steel rod has the equivalent ampacity of a 3/0 copper conductor (200 amperes@75°C). And, a 6 AWG copper conductor has an ampacity of 65 amperes @75°C. So, this appears to be a mismatch at first glance. But, the resistivity of the soil surrounding the ground rod will limit the flow of current into the earth anyway. And the fusing or melting current of a 6 AWG copper conductor for short time periods is extremely high. By far, the highest current flow through this conductor will be lightning current, which reaches a peak in about 2-10 microseconds.

The 6 AWG copper conductor has a $^1/_8$ cycle (.002 seconds) fusing current of 81,050 amperes. Installing a large conductor for connection to a ground rod(s) is certainly not necessary.

Section 250.66(A) indicates that a 4 AWG aluminum conductor may be used. 250.53(G) requires that the upper end of the rod or pipe must be flush with or below ground level, unless physical protection in the form of wood or metal is provided (250.10). And 250.64(A) prohibits aluminum or copper-clad aluminum conductors in direct contract with masonry or the earth, and, not terminated within 18″ (450mm) of the earth. So, the use of these conductors for connection to a ground rod is not permitted.

Connections to concrete-encased electrodes (250.52(B)) may be accomplished through the use of a 4 AWG copper conductor. A section of reinforcing rod may be extended externally from the concrete footing or foundation (above grade) for connection of the grounding electrode conductor. The connecting device must be listed and identified for a copper to steel connection, or exothermically welded.

Connections to ground rings (250.66(C)) must be in accordance with 250.52(A)(4), that is, through the use of a 2 AWG or larger copper conductor. The actual size of this conductor is based on the physical size of the conductor used for the ground ring. For example, 1/0 AWG copper is commonly used for the ground ring.

For grounding electrodes consisting of metal-in-ground support structures or metal underground water pipes, the sizes of the grounding electrode conductors are based on Table 250.66.

It must be noted that the grounding electrode conductor sizes for these grounding electrode types, that is, rod, pipe, or plate, concrete-encased electrode, or ground ring may be larger than specified here.

For example, if a grounding electrode conductor extended to a metal underground water pipe is sized from Table 250.66, and this conductor is extended to a ground rod(s) as a supplement to this water pipe to satisfy 250.53(D)(2), the grounding electrode conductor to the ground rod will be the same size as that required for the water pipe from Table 250.66, and not the 6 AWG copper conductor from 250.66(A),(250.64(F)).

The grounding electrode conductor sizes expressed in 250.66 should be more than adequate to perform their most important function, that is, to hold equipment and systems, at or near, earth potential ('0' volts).

It is somewhat of a belief that these conductor sizes are based on a length of 100 feet (30m). And, for lengths beyond 100 feet, the grounding electrode conductor should be increased in size to compensate for voltage-drop. This may be done by using the resistance or impedance values of Tables 8 and 9 of Chapter 9.

For example, if the grounding electrode to be used is a ground rod(s), and the grounding electrode conductor is the 6 AWG copper conductor from 250.66(A), and this conductor were to be extended 150 feet (45meters) to the ground rod, the 6 AWG copper conductor would be increased in size due to voltage-drop. Using a coated copper conductor for corrosion protection, based on the parameters given in Table 8 of Chapter 9, the 1000 foot (300m) resistance is 0.510 ohms. The 100 foot DC resistance of this conductor would be 0.051 ohms. So, the grounding electrode conductor should have a DC resistance of no more than 0.051 ohms for the 150 foot length. A 4 AWG copper conductor has a 1000 foot DC resistance of 0.321 ohms (coated). And, the 100 foot DC resistance is 0.0321 ohms. So, 0.0321×1.5 (150 feet) = 0.04815 ohms.

The grounding electrode conductor for the ground rod is 4 AWG coated copper.

While I agree that this approach seems to satisfy the intent of limiting the voltage-drop in this conductor, and, therefore, the voltage-rise above earth potential on equipment and systems, this reasoning would certainly be effective if we were dealing with 50/60 Hz currents. But, lightning currents have a frequency ranging from 3 kHz to 10 mHz. And, because inductive reactance is relative to frequency ($X_L = 2\pi FL$), it is far more important to limit the length of grounding electrode conductors, because of the increased frequency of lightning and its effects on inductive reactance, and the overall impedance of this conductor. This is the reason that 250.4(A), Informational Note No. 1 states that grounding electrode conductors are not any longer than necessary to complete the connection to the grounding electrode.

*Hermetically Sealed (as applied to Hazardous (Classified) Locations)-

Equipment sealed against the entrance of an external atmosphere where the seal is made by fusion, for example, soldering, brazing, welding, or the fusion of glass to metal. Hermetically-sealed equipment is permitted in Class I, Division 2, Class II, Division 2, and Class III, Division 1 or 2 Locations (500.7(J)).

*Identified (as applied to equipment)-

The identification of equipment for a specific purpose, function, use, or application, is typically through an organization that is acceptable to the 'Authority Having Jurisdiction'. Usually, this would be a qualified testing laboratory, such as UL, ETL, or CSA.

Sometimes, the testing and evaluation of equipment and materials is in conjunction with the American National Standards Institute (ANSI).

Some examples would be UL 489, UL 1066, and ANSI C37 for molded-case and insulated-case circuit breakers, and UL 248.1 for low-voltage fuses. (Informative Annex A).

Section 110.3(B) states that <u>listed</u> or <u>labeled</u> equipment must be used in accordance with any instructions that are included in the listing or labeling. Therefore, the designer and/or installer must be familiar with these listing instructions. I think that most of us who have been involved with electrical installations over the past 30-40 years saw a misapplication of a listing instruction which related to the 60-75°C. temperature limits of the terminals of various equipment (panelboards, disconnect switches, circuit breakers, etc.). 90°C. temperature rated insulation began to appear in 1965. And, it was common to load a conductor to its full 90° C. ampacity, without considering its 60° or 75° C. ampacity at the terminals of this equipment. References to the lower temperature limit of terminals (110.14 (C)(1)) did not appear until much later. But this listing instruction had been identified in listing directories (i.e. - UL 'Green Book' and 'White Book') long before it was referenced in Article 110.

Another example of a listing instruction that I personally missed on several occasions was the conductor fill limit of a conduit seal in a Hazardous (Classified) Location.

I assumed that the conductor fill for the conduit seal was 40% (for more than 2 conductors) (Table 1, Chapter 9), the same as typical conduit fill. But the listing instruction in the UL 'Hazardous Locations Materials Directory ('Red Book'), was identified as 25% conductor fill (501.15(C)(6)). And, once again, this listing instruction did not appear in Article 501 for many years after it appeared in the UL 'Red Book'. A larger fill percentage may be identified on the conduit seal, or a larger seal may be used through the use of reducing bushings (501.15(A)(1)).

There is also a direct correlation between Section 110.2 and 110.3 with regard to the 'Approval' of equipment and materials, and the Examination, Identification, Installation, and Use of Equipment and Materials.

The terms 'Labeled' and 'Listed' as defined in Article 100 should also be analyzed. A label is affixed to equipment and materials by an organization that is responsible for product evaluation. This label would identify specific information relating to the testing of the product and its intended use, including any information that may limit its use in certain environments. Also, see Section 90.7, and Informative Annex A, which identifies Product Standard Names with corresponding Product Standard Numbers. It is through the use of these Product Standards that appropriate listing information can be found and properly applied to satisfy Section 110.3(B). The NEC does not require that electrical equipment and materials have third party certification by an independent testing laboratory. But OSHA does require third party certification as a means of providing an additional level of safety for personnel using electrical equipment and material in accordance with OSHA Subpart S, 1910.303(a), (29 CFR, Part 1910, Subpart S).

*Information Technology Equipment-

Equipment and systems rated 1000 volts or less, normally found in offices or other business establishments and similar environments classified as ordinary locations, that are used for creation and manipulation of data, voice, video, and similar signals that are not communications equipment as defined in Part I of Article 100, and do not process communications circuits as defined in 800.2 (NFPA 175-UL 60950-1-2014). Article 645 has specific requirements relating to this type of equipment and to the electrical supply circuits and interconnecting cables, electrical supply cords, data cables, and equipment grounding conductors under a raised floor, UPS equipment, system grounding, equipment grounding and bonding, and surge protection for Critical Operations Data Systems and Selective Coordination for these critical systems.

Branch circuit conductors that supply one or more units of Information Technology Equipment are required to have an ampacity of 125% of   the load to be served (645.5(A)).

Power supply cords are limited to a length of 15 feet (4.5 meters) and they must be listed for Information Technology Equipment (645.5(B)). Interconnecting cables are not limited to a 15 foot length (4.5 meters), and these cables are required to be listed for this type of equipment (645.5(C)).

The wiring methods under a raised floor are referenced in 645.5(E)(1),(a).

*Innerduct-

A nonmetallic raceway placed within a larger raceway. An example of this use would be from Table 800.154(b), where a listed plenum communications raceway, a listed riser communications raceway, or a listed general-purpose communications raceway is installed within any of the listed raceway types for Optical Fiber Cables in Chapter 3 (770.110(A)(3)).

*In Sight From, Within Sight From, Within Sight-

Generally, this means that one equipment is visible, and not more than 50 feet (15 m) distant from another equipment. Section 430.102(A), as it relates to the location of the disconnecting means for a motor controller, would be the most common use of this term. However, there are certain exceptions. For motor circuits of over 1000 volts, the disconnect (430.102(A), Exception, No.1) is not required to be 'in sight from' the controller, providing that it is lockable and the controller has a warning label which indicates the location of the disconnect. Also, Exceptions are provided for a group of coordinated controllers on a multimotor process machine, where a single disconnecting means is provided for this group of controllers, instead of each controller having its own disconnect. In this case, both the disconnect and the group of controllers must be 'in sight from' the machine.

Another Exception that waives this requirement is for 'valve actuator motor assemblies', where the disconnecting means is lockable.

Section 440.13 permits an attachment plug and receptacle to serve as the disconnecting means for a room air conditioner, household refrigerators and freezers, drinking water colors, and beverage dispensers.

For air conditioning or refrigerating equipment, the disconnecting means is also to be 'in sight from' and 'readily accessible' from this equipment (440.14).

For room air conditioners with voltages up to 250 volts, the plug and receptacle may serve this purpose, if the manual controls for the unit are readily accessible and within 6 feet (1.8m) from the floor. Otherwise, a manually operable disconnecting means is required and it must be readily accessible and 'in sight from' the room air conditioner (440.63).

Also, see Section 600.6(A)(3) for these requirements for signs and outline lighting.

*Interactive Inverter-

An inverter intended for use in parallel with an electric utility to supply common loads that may deliver power to the utility.

An interactive system is defined as a photovoltaic system that operates in parallel with, and may deliver power to an electrical production and distribution network. The electrical production and distribution network is usually a utility which is not controlled by the PV power system (690.2).

*Interrupting Rating-

'The highest current at rated voltage that a device is intended to interrupt under standard test conditions'. This term and definition are associated with a circuit breaker, and its ability to operate under short-circuit or ground-fault conditions without exploding or rupturing in the process. This is also related to the 'interrupting capacity' of fuses. And there is a direct relationship between this defined term and Section 110.9. Circuit breakers have interrupting ratings ranging from 5000 amperes to 200,000 amperes, symmetrical. Fuses have interrupting capacities ranging from 10,000 amperes to 300,000 amperes, symmetrical. And the fault-clearing times of these devices may be as low as 1/4 cycle (.004 seconds) for current-limiting fuses. In order to satisfy the provisions of Section 110.9, the available fault current must be determined at the line terminals of the overcurrent device, and the overcurrent device must be able to interrupt this fault current. Section 240.83(C) states that where the interrupting rating of a circuit breaker is other than 5000 amperes, the interrupting rating must be identified on the circuit breaker.

Of course, determining the available short-circuit current at different places in an electrical distribution system is also necessary to establish the short-circuit withstand rating of equipment. When properly calculated and applied, this will assure that the protective system and downstream equipment will meet the requirements of Section 110.10.

Interrupting fault current is not the only consideration. A motor circuit switch, rated in horsepower (430.109(A), and this also applies to motor controllers with identified horsepower ratings (430.83(A)(1)), will serve to interrupt the locked-rotor current of the motor. Equivalent locked-rotor currents are listed in Tables 430.251(A) and 430.251(B).

And the motor horsepower ratings also play a role in the ampere rating of a motor disconnect switch for combination loads. When supplying loads consisting of two or more motors, or where one or more motors are used in conjunction with other loads, the rating of the motor circuit switch is determined by adding the currents of all motors at the full-load condition and the locked-rotor condition to the ampere rating of all other loads. And, this total current is to be considered as a single motor in the selection of the rating of the motor circuit switch. The combined full-load current of all of the load is increased by a factor of 1.15 to determine the ampere rating of the combined load (430.110(C)(1), (430.110(C)(2)).

For example, a three-phase 480 volt (460 volt, nominal) system supplies 3 motors rated at 7½ HP, 5 HP, and 3 HP. All motors are Design B, and, in addition, a 10 kW electric heating load is supplied. From Table 430.250, the motors are rated - 11 amperes, 7.6 amperes, and 4.8 amperes, respectively. The heating load has a rating of 12 amperes.

Ampere Rating

$$\frac{10,000\,Watts}{480\,V.\times1.732} = 12\,amperes$$

Total Load 11.0 amperes
7.6 amperes
4.8 amperes
12.0 amperes - Heating Load
35.4 amperes

35.4 amperes
× 1.15
40.71 amperes

A standard 60 ampere-rated switch would be acceptable.

Horsepower Rating

Table 430.251 (B)

3 HP -	*32.0 amperes*
5 HP -	*46.0 amperes*
7 1/2 HP-	*63.5 amperes*
	141.5 amperes
	12.0 amperes -Heating Load
	153.5 amperes

Minimum horsepower rating of switch from Table 430.251(B) is 25 HP (183 amperes).

Note: There are exceptions to 430.110(A) and 430.110(C) that waive the 115% value for the switch ampere rating for 'listed' nonfused motor-circuit switches that have horsepower ratings not less than the motor horsepower rating for single motors, or the combined horsepower ratings for multiple motors applications.

The interrupting ratings and ampere ratings of the disconnecting means for Hermetic Refrigerant Motor Compressors are calculated using a similar method as for typical motors (440.12(A)(1)), (440.12(A)(2)), (440.12 B)(l)), (440.12(B)(2)). One difference is that the calculations are based on the nameplate rated-load current or the branch-circuit selection current (if specified on the nameplate), whichever is greater.

Once again, this is to establish the ampere rating and horsepower rating of the disconnecting means, whether it is a single motor, or a combination load.

*Intersystem Bonding-

In order to reduce the effects of voltage differences between different systems, it is imperative that the grounding electrodes of these systems, including the service supplied system, the communications system, any separately-derived system, and the lightning protection system, be bonded together to form one grounding electrode system. This has been already covered, in detail, in our analysis of the terms in Article 100 associated with proper grounding methods. Sections 250.92 and 250.94 address these bonding requirements. Also, Sections 250.60 and 250.106 covers the required bonding of the grounding electrode(s) for a lightning protection system.

In Chapters 7 and 8, we can find the references for the bonding of the grounding electrodes for Optical Fiber Cable installations (770.100(D)), Communications Systems (800.100(D)), Radio and Television Equipment (810.21(J)), CATV and Radio Distribution Systems (820.100(D)), Network-Powered Broadband Communications Systems (830.100(D)) and Premises Powered Broadband Communications Systems (840.106(B)).

Once again, the bonding of these grounding electrodes will serve to reduce, but not completely eliminate, the voltage differences that may appear due to the effects of power faults or lightning.

*Intrinsically-Safe Apparatus-

Apparatus in which all of the circuits are intrinsically-safe. Intrinsically-safe circuits are incapable of producing ignition capable energy during normal or abnormal conditions (504.2), such as two simultaneous faults, which include component failures and overcurrent conditions.

*Intrinsically-Safe System (as applied to Hazardous (Classified) Locations-

An assembly of interconnected intrinsically-safe apparatus and interconnected cables, in that those parts of the system that may be used in Hazardous (Classified) Locations are intrinsically-safe circuits. This system may include a single circuit or multiple circuits that have been evaluated and determined to be intrinsically-safe.

A common protection technique for instrumentation circuits is a zener-diode barrier, where the zener-diodes are installed in parallel with the supply circuit in a redundant mode, with the return circuit providing a low-impedance path to ground. The intrinsically safe ground bus within the zener-diode barrier (in the nonhazardous location) must be connected to ground (earth), and the resistance between the IT ground bus and the earth ground must be less than 1 ohm. This will limit the voltage-rise on the instrument circuit during fault conditions. The input voltage will be less than that of the diodes, with the redundant-diode designed to conduct at 1-2 volts above the first diode.

There will be 2 resistors in the supply circuit, which along with a fuse and the zener-diodes, will limit the energy in the hazardous location to below the ignition energy of the specific gas or vapor in the Class I environment (UL 913).

*Kitchen-

An area with a sink and permanent provisions for food preparation and cooking. Sections 220.55 and 220.56 have extensive information on the subject of calculating electrical loads in dwelling and nondwelling occupancies, including demand factors for kitchen equipment.

Section 220.52(A) is a reference for the small appliance load in a dwelling unit kitchen, identified as the 'small appliance load' (210.11(C)(1), (210.52 (B)(1)). There must be a minimum of 2-20 ampere branch circuits for receptacle outlets in the kitchen and dining areas.

Section 210.8(A)(6) requires GFCI protection for receptacles that serve countertop surfaces in dwelling units.

Section 210.8(B)(2) specifies that all 15 and 20 ampere receptacles in nondwelling kitchens are to be GFCI protected.

Section 210.8(D) requires GFCI protection for a dishwasher in dwelling units.

*Mobile Equipment-

Equipment with electrical components suitable to be moved only with mechanical aids, or is provided with wheels for movement by a person(s) or powered devices.

*Motor Control Center-

This is a cabinet which contains motor controllers (starters), with each controller connected to a common power bus. Part VIII of Article 430 applies to Motor Control Centers (Sections 430.92, 430.94, 430.95, and 430.96). Overcurrent protection is in accordance with Article 240, and for the individual motor circuits, in accordance with Part IV of Article 430.The grounding provisions of Article 250 that involve equipment grounding (Part VI of Article 250) would also apply. And, the available short-circuit current at the Motor Control Center must be calculated, and the date that the calculation was performed must be documented and made available to those authorized to inspect the installation (430.99).

*Neutral Conductor-

This conductor attaches to the neutral point of a system. The neutral conductor is solidly grounded at the service equipment, and, also at the neutral terminal of the supply transformer. Or, at the source of a separately-derived system. In lieu of the connection at the source, the neutral may be solidly grounded at the first system disconnect or overcurrent device supplied by the separately-derived system (Part II - Article 250).

In addition, the Neutral Conductor is subject to the installation requirements of Section 200.4, and the identification requirements of Sections 200.6, 200.7, 200.9, and 200.10.

The neutral conductor is a 'grounded conductor' because it is intentionally grounded. However, not all 'grounded conductors' are 'neutral' conductors, such as a 'corner-grounded delta system', where one phase is intentionally grounded.

*Neutral Point-

The 'neutral point' of a system is the point where voltages from all other points are equal. This would include a 3-phase, Wye connected system, where the voltages from XI, X2, and X3 would be the same when measured to the neutral point (X0). Or, the midpoint of a single-phase system, where there would be an equal voltage to this midpoint. Also, the midpoint of the single-phase winding of a 3-phase Delta system and the midpoint of a 3-wire direct-current system establish the neutral point.

*Nonconductive Optical Fiber Cable-

A factory assembly of one or more optical fibers having an overall covering and containing no electrically conductive materials. Optical Fiber Cable, whether it contains conducting members or not, is required to be listed and identified in accordance with 770.179(A) through (F). An Exception to this listing requirement appears in 770.48(A),(B)).

Unlisted Optical Fiber Cable is permitted to be installed within building spaces where this cable enters the building through an external wall or floor slab and the length does not exceed 50 feet (15 meters) and the cable terminates in an enclosure.

Unlisted Nonconductive Outside Plant Optical Fiber Cable is permitted to enter the building from the outside through an exterior wall or floor slab where installed in intermediate metal conduit, rigid metal conduit, rigid polyvinyl chloride conduit, or electrical metallic tubing. However, this type of cable is not permitted to be installed rigid polyvinyl chloride conduit and electrical metallic tubing in risers, ducts for environmental air, plenums used for environmental air, and other spaces used for environmental air.

*Nonincendive Circuit (As applied to Hazardous (Classified) Locations)-

A circuit, other than field wiring, in which any arc or thermal effect produced under intended operating conditions of the equipment, is not capable, under specified test conditions of igniting the flammable gas-air, vapor-air, or dust-air mixture (ANSI/ISA 12.12.01-2013).

*Nonlinear Load-

A linear load is one where the wave shape of the steady-state current follows the wave shape of the applied voltage. Examples of this type of load include incandescent lighting and electric resistance heaters. This type of load produces voltage and current waveforms that are 'sinusoidal' and the waveforms of the voltage and current are said to be 'in-phase' with one another. The circuit impedance remains constant, even as the voltage changes. No change of frequency is registered and 'harmonics' are not produced.

However, other types of load may cause a distorted wave-form to be produced. Electronic equipment, such as computers, monitors, and copiers, and other types of equipment with 'Switch-Mode Power Supplies', and electric-discharge lighting systems cause a distortion of the current waveform and this produces a distortion of the voltage waveform. In difference to a linear load, where the circuit impedance does not change with a change in applied voltage, in a nonlinear load, the impedance of the circuit does change as the applied voltage changes. And, this produces a distorted waveform (nonsinusoidal). So, the waveform of the steady-state current no longer follows the waveshape of the applied voltage.

The harmonic currents produced by these nonlinear loads may range from the third (triplen) to beyond the twenty-first harmonic.

It is the third harmonic current (180 cycle) that may become a problem where nonlinear loads are supplied by a 3-phase, 4-wire, Wye connected distribution system. In this case, the triplen harmonic current from the 3-phase conductors will not be effectively cancelled, as would be the case for normal 60 cycle currents. Instead, these harmonic currents add, vectorially, and flow though the neutral conductor. Depending on the amount of load being supplied on this 3-phase system, the neutral conductor may become seriously overloaded. Theoretically, the neutral conductor may carry as much as 3 times the current flowing in the phase conductors, but, in reality, this current may be as much as 2 times the current in the 3-phase conductors. This is a 180 cycle (third harmonic) current, and, if the neutral conductor were not increased in size to accommodate this excessive current, dangerous overheating would be the result. In addition, the supply transformer would be subject to overheating, and failure would eventually occur. And this condition would remain a problem, even if the electrical load was virtually balanced.

Ironically, this additive current in the neutral of a 3-phase, 4-wire, Wye connected distribution system is not a problem if the same type of nonlinear load was supplied from a single-phase system (120/240 volt). The harmonic currents produced in the 2 ungrounded conductors would be effectively cancelled, and they would not add in the neutral conductor.

The NEC is not a design specification (90.1(A)). So, there are no recommendations on how to correct an overheating problem associated with harmonic currents. Doubling the cross-sectional area of the neutral conductor, and increasing the capacity of the supply transformer, using 'K' rated transformers, as well as installing a harmonic filter to correct the excessive harmonic currents are certainly options that should be considered.

Section 310.15(A)(3), Informational Note No. 1, which provides some guidance on the topic of the temperature rating of a conductor, identifies the four principal determining factors associated with conductor operating temperature. The second of these four factors to consider is the actual load on the conductor, including <u>fundamental (60 Hz) and harmonic currents.</u> The fourth determining factor is the effect of adjacent current-carrying conductors. Section 310.15(B)(5),(c) states that on a 3-phase, 4-wire, Wye system, where a major portion of the load is nonlinear, the neutral conductor is considered to be a current-carrying conductor due to the presence of harmonic currents. This would necessitate the application of an 80% ampacity adjustment factor from Table 310.15(B)(3),(a).

I have heard about a suggestion to use a zigzag autotransformer to reduce or eliminate the triplen harmonic currents associated with nonlinear loads.

The use of an autotransformer to create a grounded neutral conductor for the purpose of supplying nonlinear loads from a 480/277 volt or 208/120 volt distribution system is a viable method of reducing or eliminating the harmonic currents that are associated with this type of load.

The autotransformer would be installed in close proximity to the nonlinear load. The harmonic currents produced by the nonlinear load will not pass through the windings of the autotransformer, and, therefore will not be introduced into the upstream supply system, where dangerous overheating of the supply conductors and transformer may be the result.

The neutral point of the autotransformer would be connected to the building grounding electrode system.

And, this is where we have a problem, and the reason that this type of installation is not acceptable, and it is prohibited by NEC 450.5 and 215.11.

Section 450.5 only recognizes the use of this transformer on 3-phase, 3-wire <u>ungrounded systems.</u>

If the supply system is solidly grounded and the zigzag autotransformer is used to create a neutral point for downstream single-phase, line-to-neutral connected loads, and a ground-fault occurs on the upstream supply system, the ground-fault current will be shared by the supply system transformer and the zigzag autotransformer. The supply system transformer is typically of a larger size than the zigzag autotransformer, and serious damage to the autotransformer will be the result.

*Outlet-

This is defined as a point on the wiring system where current supplies 'utilization equipment'. The outlet may be 'hard-wired' or designed for a 'cord-and-plug' connection, regardless of the voltage (210.50),(210.52), (210.63), (220.14).

Article 210-Part III identifies the outlets that are required.

Article 314 is the most commonly used reference associated with the term 'outlet'.

When sizing outlet, device, junction boxes, and conduit bodies, the provisions of 314.16 apply.

The volume of the box is the total volume of the assembled sections, which may include extension rings, domed covers, and plaster rings, which are identified with their marked volume.

The boxes may be assembled from those identified in Table 314.16(A).

If the box has one or more securely installed barriers, a volume deduction must be made in accordance with the marked volume on the barrier, or ½

Table 314.16(A) Metal Boxes

Box Trade Size			Minimum volume		Maximum number of conductors* (arranged by AWG size)					
mm	in.		cm³	in.³	18	16	14	12	10	8
100 × 32	(4 × 1¼)	round/octagonal	205	12.5	8	7	6	5	5	5
100 × 38	(4 × 1½)	round/octagonal	254	15.5	10	8	7	6	6	5
100 × 54	(4 × 2¹/₈)	round/octagonal	353	21.5	14	12	10	9	8	7
100 × 32	(4 × 1¼)	Square	295	18.0	12	10	9	8	7	6
100 × 38	(4 × 1½)	Square	344	21.0	14	12	10	9	8	7
100 × 54	(4 × 2¹/₈)	Square	497	30.3	20	17	15	13	12	10
120 × 32	(4¹¹/₁₆ × 1¼)	Square	418	25.5	17	14	12	11	10	8
120 × 38	(4¹¹/₁₆ × 1½)	Square	484	29.5	19	16	14	13	11	9
120 × 54	(4¹¹/₁₆ × 2¹/₈)	Square	689	42.0	28	24	21	18	16	14
75 × 50 × 38	(3 × 2 × 1½)	device	123	7.5	5	4	3	3	3	2
75 × 50 × 50	(3 × 2 × 2)	device	164	10.0	6	5	5	4	4	3
75 × 50 × 57	(3 × 2 × 2¼)	device	172	10.5	7	6	5	4	4	3
75 × 50 × 65	(3 × 2 × 2½)	device	205	12.5	8	7	6	5	5	4
75 × 50 × 70	(3 × 2 × 2¾)	device	230	14.0	9	8	7	6	5	4
75 × 50 × 90	(3 × 2 × 3½)	device	295	18.0	12	10	9	8	7	6
100 × 54 × 38	(4 × 2¹/₈ × 1½)	device	169	10.3	6	5	5	4	4	3
100 × 54 × 48	(4 × 2¼ × 1⁷/₈)	device	213	13.0	8	7	6	5	5	4
100 × 54 × 54	(4 × 2¹/₈ × 2¹/₈)	device	238	14.5	9	8	7	6	5	4
95 × 50 × 65	(3¾ × 2 × 2½)	masonry box	230	14.0	9	8	7	6	5	4
95 × 50 × 90	(3¾ × 2 × 3½)	masonry box	344	21.0	14	12	10	9	8	7
min. 44.5 depth	FS – single cover (1¾)		221	13.5	9	7	6	6	5	4
min. 60.3 depth	FD – single cover (2³/₈)		295	18.0	12	10	9	8	7	6
min. 44.5 depth	FS – multiple cover (1¾)		295	18.0	12	10	9	8	7	6
min. 60.3 depth	FD – multiple cover (2³/₈)		395	24.0	16	13	12	10	9	8

Table 314.16(B) Volume Allowance Required per Conductor

Size of conductor (AWG)	Free Space within Box for each Conductor	
	cm^3	in.^{3}
18	24.6	1.50
16	28.7	1.75
14	32.8	2.00
12	36.9	2.25
10	41.0	2.50
8	49.2	3.00
6	81.9	5.00

cubic inch (8.2 cubic centimeters) if, metal, or 1 cubic inch (16.4 cubic centimeters), if nonmetallic.

Each space where barriers are provided is calculated separately.

When determining conductor fill, each conductor originating outside of the box and terminating inside the box is counted as one conductor, and each conductor that passes through the box without being spliced or terminated is counted as one conductor. Where a conductor is looped or coiled and not broken, with a length of not less than twice the minimum length required for free conductors in 300.14, the conductor is counted twice. The length of free conductor, measured from the point of emergence from a raceway or cable sheath must be a minimum of 6 inches (150mm).

Conductor fill is based on the volume allowance of Table 314.16(B). A conductor that originates within the box and does not leave the box, for example, a jumper wire, is not counted when making volume deductions.

An equipment grounding conductor(s), or not over 4 fixture wires smaller than 14 AWG, or both, are not counted for volume deductions where they enter a box from a domed luminaire or similar canopy.

However, equipment grounding conductors or bonding jumpers that enter a box count as a single conductor that is based on the largest equipment grounding conductor or bonding jumper within the box, based on the volume allowance specified in Table 314.16(B). If one or more equipment grounding conductors are installed for isolated grounding-type receptacles (250.146(D)), in addition to the equipment grounding conductors for other circuits, an additional conductor, based on the largest equipment grounding conductor for the isolated grounding type receptacle(s), must be added when calculating the required box dimension.

Internal cable clamps count as a single conductor, based on the largest conductor within the box.

For one or more luminaire studs or hickeys present in the box, a single volume deduction from Table 314.16(B), for each type of fitting must be made, based on the largest conductor within the box.

For each yoke or strap containing one or more devices or equipment, a two wire deduction, based on Table 314.16(B), must be made, in accordance with the largest size conductor connected to the device or equipment.

However, if the device or equipment is wider than 2 inches (50mm), a double volume deduction must be made for each gang required for mounting.

Example

What is the required volume of a 4 inch square outlet box containing a single-pole switch and a duplex receptacle, where the switch is supplied by a 14/2, with ground, NM cable, and the duplex receptacle is supplied by a 12/3, with ground, NM cable?

Volume Deductions

14/2 NM cable – 14AWG – 2 cubic inches each – 4 cubic inches
12/3 NM cable – 12AWG – 2.25 cubic inches each – 6.75 cubic inches
2 internal cable clamps – 1 volume deduction based on a 12AWG conductor – 2.25 cu.in.
2 equipment grounding conductors – 1 volume deduction based on a 12AWG conductor – 2.25 cu.in.
1 single-pole switch –2-14AWG conductors– 4 cubic inches
1 duplex receptacle –2-12 AWG conductors – 4.5cu.in.

$$4\,cubic\,inches$$
$$6.75\,cubic\,inches$$
$$2.25\,cubic\,inches$$
$$2.25\,cubic\,inches$$
$$4\,cubic\,inches$$
$$\underline{4.5\,cubic\,inches}$$
$$23.75\,cubic\,inches$$

Answer - $4'' \times 2^{1}/_{8}''$ square box – 30.3 cu.in. - Table 314.16(A)

For raceways or cables containing conductors 4 AWG or larger, the minimum dimensions of pull or junction boxes are in accordance with 314.28.

For straight pulls, the length of the box is based on 8 times the trade diameter of the largest raceway.

For example, where the largest raceway size is 3 inch, the length of the box would be a minimum of 8×3, or 24 inches. This would apply where the conductors are run through the box without joints or splices.

If the conductors are spliced in the box, the length of the box may be 6 times the trade diameter of the largest raceway, or 18 inches in this example.

For angle or U pulls, or where splices are made, the box or conduit body dimension is based on at least 6 times the trade diameter of the largest raceway size in a row (if there is more than 1 row). This dimension is increased for additional conduit entries on the same wall of the box. If there is more than 1 row, each row is calculated individually, and the row that provides the maximum distance is to be used.

For example, where 3 conduits enter and leave a box at right angles, and their sizes are 3 inch, 2 inch, and 1½ inch, the box dimension would be:

$$6 \times 3 = 18''+2'' + 1\frac{1}{2}'' = \underline{\mathbf{21\frac{1}{2}''}} \times \underline{\mathbf{21\frac{1}{2}''}}$$

Additional terms regarding the definition of 'Outlet' are as follows:

Cord Pendants

A cord connector that is supplied by a permanently connected cord pendant is considered to be a receptacle outlet.

Dwelling Unit Receptacle Outlets

Appliance Outlets

Receptacle outlets in a dwelling occupancy for specific appliances are to be within 6 feet (1.8m) of the appliance location (210.50(C)).

Section 210.52 covers the requirements for dwelling unit receptacle outlets. These locations include kitchens, family rooms, dining rooms, living rooms, parlors, libraries, dens, sunrooms, bedrooms, and recreation rooms.

The spacing of the receptacles must assure that no point measured horizontally along the floor line of any wall space is more than 6 feet (1.8m) from a receptacle outlet (210.52((A)(1)).

The wall space includes any space 2 feet (600mm) or more in width and unbroken by doorways, or similar work surfaces, fixed panels in walls, except sliding panels, and fixed-room dividers, such as free-standing bar-type counters or railings (210.52((A)(2)).

Floors receptacles are not counted as part of the required number of receptacles, unless they are within 18 inches (450mm) of the wall (210.52(A)(3)).

Countertop receptacles, such as in a kitchen, are not counted to satisfy the requirements in 210.52(A)(4).

At least one receptacle outlet is required for a wall countertop and work surface that is 12 inches (300mm), or wider. And, no point along the wall line is more than 24 inches (600mm), measured horizontally, from a receptacle outlet in that space (210.52(C)(1)). This does not require a receptacle outlet on a wall behind a range, counter-mounted cooking unit, or sink.

Island countertops require at least one receptacle where the long dimension is 24 inches (600mm) or greater and the short dimension is 12 inches (300mm) or greater. This also applies to peninsular countertop spaces. However, the peninsular countertop dimension is measured from the connected perpendicular wall. This means that a receptacle outlet in this wall may satisfy this requirement, depending on the peninsular countertop dimension (210.52(C)(2)).

Receptacle outlets are to be located on or above the countertop or work surface, but not more than 20 inches (500mm) above (210.52(C)(5)).

Where the countertop spaces are separated by range tops, refrigerators, or sinks, these spaces are considered separate countertop spaces (210.52(C)(4)).

At least one receptacle outlet must be installed in bathrooms, and within 3 feet (900mm) of the outside edge of each basin. This receptacle may be located on a wall or partition adjacent to the basin countertop, or on the countertop, or on the side of the basin cabinet, but not more than 12 inches (300mm) below the top of the basin or basin countertop. If the receptacle outlet assembly is listed to be installed in the countertop space, in accordance with 406.5(E),(G), this is acceptable if the receptacle is within the 3 foot (900mm) dimension from the outside edge of the basin (210.52(D)).

For one and two family dwellings, and for each unit of a two family dwelling that is at grade level, there must be at least one receptacle, not more than 6½ feet (2m) above grade level at the front and back of the dwelling (210.52(E)(1)).

For multifamily dwellings, each unit at grade level and provided with individual exterior entrance and egress must have a readily accessible receptacle outlet, which is not more than 6½ feet (2m) above grade level (210.52(E)(2)).

Where there are balconies, decks, and porches attached to the dwelling and they are accessible from the inside, at least one receptacle outlet which is accessible from the balcony, deck, or porch is required not more than 6½ feet (2.0m) above the balcony, deck, or porch (210.52(E)(3)).

For laundry areas in dwelling units, there must be at least one receptacle outlet (210.52(F)), and this receptacle outlet(s) must be supplied by a dedicated 20 ampere branch that supplies no other outlets (210.11(C)(2)).

For attached garages and detached garages with electric power, there must be at least one receptacle outlet in each vehicle bay, not more the 5½ feet (1.7m) above the floor (210.52(G)(1)).

In each separate unfinished portion of a basement, at least one receptacle outlet is required (210.52(G)(3)).

In each accessory building with electric power, at least one receptacle outlet is required (210.52(G)(2)).

*Overcurrent-

An overcurrent condition is one that is caused by an overload, short-circuit, or ground-fault. An overcurrent device, typically in the form of a fuse or circuit breaker, provides protection for any of these conditions. Article 240 is the primary source of information on this topic. Section 240.4 has extensive information on the overcurrent protection of conductors. And Section 240.21 applies to various 'tap rules', where smaller conductors may be connected or 'tapped' to larger conductors. These smaller conductors are considered to be protected by the overcurrent device protecting the larger conductors.

The 10-foot tap rule would permit these smaller 'tap conductors' to be protected at up to 10 times their normal ampacity (240.21(B)(1)). This would mean, for example, that a 10 AWG THHN insulated copper conductor (5.26mm^2), which has a normal ampacity of 40 amperes (90°C.), may be tapped to a larger feeder conductor, which is protected at 400 amperes. However, caution must be exercised here because this 400 ampere protection, fuse or circuit breaker, must still provide short-circuit and ground-fault protection for the 10 AWG THHN copper conductor. From our previous discussions on overcurrent protection of conductors, we must determine the available fault-current that the 10 AWG THHN copper conductor may be subjected to through a short-circuit analysis. And then determine whether the 400 ampere overcurrent device protecting the larger feeder conductors will also provide short-circuit and ground-fault protection for the smaller tap conductors in accordance with their short-time insulation withstand ratings. So, a lot of careful study would be necessary before using this, or any other tap rule.

For overcurrent protection of motor branch circuits, Sections 430.51 through 58 would apply. For motor feeders, Sections 430.61, 430.62, and 430.63 would apply. For motor control circuits, Sections 430.71 and 430.72 would apply.

For air conditioners and refrigerating equipment, Sections 440.21 and 440.22 would apply.

For transformers, overcurrent protection for the primary and secondary (if necessary), is listed in Section 450.3. And, for autotransformers, in Section 450.4.

Information relating to overcurrent protection of Phase Converters can be found in Section 455.7.

Overcurrent protection for each ungrounded conductor of a capacitor bank is a requirement of 460.8(B). And, if adjustable, the rating or setting must be as low as practicable. No overcurrent device is required where a capacitor is on the load side of a motor overload device (460.8(B), Exception).

When we summarized the definition of the term 'Main Bonding Jumper', a three-phase, 1000 kVA, 13,800/480/277 volt transformer, with a 3.5 percent impedance, supplied the service equipment. The service conductors were a paralleled set of 3-600 kcmil copper conductors, per phase, and a full-sized neutral. And, these secondary conductors were 100 feet (30.48m) long.

The available short-circuit current at the transformer secondary was calculated at 38,195 amperes, and the calculated short-circuit current at the service disconnect, based on these conditions, was 32,817 amperes, where the service conductors were within nonmagnetic (possibly, PVC) raceways.

If a 400 ampere fusible switch were installed at the service equipment, and, for calculation purposes, a 3-phase, 4-wire, 400 ampere feeder extended a length of 150 feet (45.72 meters) to a panelboard, and the feeder conductors were 500 kcmil THHN copper (380 amperes at 75°C.) in a steel (magnetic raceway), and the 10 foot tap was made at the 100 foot (30.48m) length of this feeder, we can determine whether the 10 AWG (5.26mm²) THHN copper conductors have been provided with the proper short-circuit protection to assure compliance with 110.10.

Let's assume that the 400 ampere fuses have an interrupting capacity of 100,000 amperes symmetrical, which is well above the 32,817 amperes of short-circuit current that is available at the line terminals of the fusible switch, and this complies with 110.9.

The fault clearing time of the fuses in this example are one cycle, or .016 seconds.

$$\frac{1.732 \times 100 \times 32,817}{22,185 \times 480V} = \frac{5,683,904}{10,648,800} = .5337$$

$$\frac{1}{1+.5337} = .6520$$

$$
\begin{array}{ll}
32,817\,amperes & (at\ service\ equipment) \\
\times\ .6520 & \\
\hline
21,397\,amperes & (100\ feet\ where\ the\ tap\ is\ made)
\end{array}
$$

The insulation withstands rating of the 10 AWG THHN copper conductor is 4349 amperes for one cycle (.016 seconds).

$$10\,AWG - \frac{10.380\,circular\;mils}{(one\;ampere\;for\;every\,42.25cm\;for\,5\,seconds)} = 246\,amperes$$

246 amperes × 246 amperes × 5 seconds = 302,580 (ampere-squared seconds)

$$\frac{302,580}{.016\,(clearing\;time\,of\;fuses)} = 18,911,250$$

$$\sqrt{18,911,250} = 4349\,amperes.$$

The available short-circuit current where the 10 AWG THHN copper conductors tap to the 500 kcmil copper conductors is 21,397 amperes. And, the fault-clearing time of the 400 ampere fuses at 21,397 amperes, has been determined to be one-cycle.

However, the one-cycle insulation withstands rating of the 10 AWG tap conductor is only 4349 amperes. Certainly, the insulation on the 10 AWG copper conductor will be destroyed under these conditions. As a matter of fact, this fault-current, for one-cycle, is even well above the fusing or melting current of this conductor (11,331 amperes). So the 10 AWG conductor will be destroyed. This condition is definitely a violation of 110.10, as the tap conductors are not protected in accordance with their insulation withstand rating.

This is an example where the provisions of the 10 foot tap rule appear to be satisfied in accordance with 240.21(B)(1). But, the provisions of 110.10 have not been satisfied. Merely providing overcurrent protection that does not exceed 10 times the ampacity of the tap conductors, in this case, 400 amperes for the 10 AWG THHN copper conductors (310.15(B)(16)), is not acceptable.

Based on these circuit conditions, with an available short-circuit current of 21,397 amperes where the 10 foot tap is to be made, the minimum size copper conductor that may be used for the tap conductors is 3 AWG copper (26.66mm²). The one-cycle insulation withstand rating for this size conductor is 22,008 amperes.

And, finally, if the 400 ampere fuses are current-limiting devices, with a fault-clearing time of less than ½ cycle (.008 seconds), the 10 foot tap conductor may be 4 AWG copper, (21.15mm²), which has a ½ cycle insulation withstand rating of 24,700 amperes.

*Overcurrent Protective Device, Branch-Circuit-

This device is meant to provide protection against the damaging effects associated with overloads, short-circuits, and ground-faults. Interrupting ratings must be a minimum of 5,000 amperes (240.83(C), and may be as high as 300,000 amperes for some current-limiting overcurrent protective devices. The minimum interrupting rating (capacity) of these devices is in accordance with the available fault-current at the line terminal(s) of the device (110.9) And, based on the available fault-current and the operating characteristics of the overcurrent device (fault-clearing time), we can determine whether this device provides the proper overcurrent protection in accordance with the short-circuit withstand rating of the conductor insulation, as well as the electrical equipment on the load side of the overcurrent protective device (110.10).

*Overcurrent Protective Device, Supplementary-

This overcurrent device may be used in conjunction with an upstream branch-circuit overcurrent device, typically to provide additional protection for a piece of equipment, and, to isolate a single piece of equipment without affecting other equipment on the same branch-circuit.

Supplementary overcurrent devices are in common use for luminaires, appliances, and HVAC equipment (240.10). These devices do not have an interrupting rating, as they are not tested under fault conditions, and they are not intended to be used for branch-circuit protection in accordance with UL 1077. These devices must be on the load side of an overcurrent device that provides the overload and fault protection for the supplementary overcurrent device.

*Overload-

Fuses and circuit breakers are designed to provide protection from short-circuits, ground-faults, and overloads, within their interrupting range.

Most commonly, short-circuits will normally produce sufficient current to initiate the operation of an overcurrent device in a very short time.

Overloads, where they persist for long periods, may or may not, be cleared by the circuit overcurrent device. An inverse-time circuit breaker, at three times its rating, may take several minutes to clear.

For example, this type of circuit breaker, at 251–600 volts, and a rating of 100 amperes, will take up to 160 seconds to open at the same voltage. And, at a rating of 400 amperes, it will take up to 350 seconds to open.

While inverse-time circuit breakers lend themselves very effectively to motor circuits, where overloads during starting may be 6–8 times the normal motor load, a sustained load of 3 times the circuit breaker rating for several minutes will have an adverse effect on the service life of circuit components due to thermal stress. Increases in operating temperature of 5–10° C. may reduce the service life of conductor insulation by 50%. Terminations subjected to heating and cooling cycles will expand and contract and loosen over time, which will cause additional heating and eventual failure.

At one time, it was a common maintenance practice to periodically check the tightness of wiring connections with standard tools, (wrenches, screw drivers, and allen wrenches). The maintenance technician would, very likely, apply additional force to the termination, not realizing that, over time, as the wire strands spread and exert force on the terminal, the connection would be destroyed. Indeed, in the mid-1960's. I worked for an electrical contractor, and several customers would hire us to do this type of maintenance, usually on a yearly basis. And, they were actually paying us to damage their equipment!!!

The following procedure should be used to avoid connection problems.

Initially, when equipment is first installed, a calibrated torque measurement tool is used to tighten the screw or bolt to the manufacturer's specification. Informative Annex I has information on the recommended tightening torque from UL Standard 486 A-B (110.14(D)).

A very important recommendation, and an excellent practice, is to mark a straight line across the screw or bolt and the stationary part of the termination. A visual inspection will identify whether the screw or bolt has moved after the proper torque has been applied, as some relaxation of the connection over time is normal.

Another industry practice is to use the calibrated torque measurement tool to check existing terminations at 90% of the specified torque value. If the screw or bolt does not move, the termination is properly torqued.

If the screw or bolt does move, this indicates that the termination should be reinstalled (**NFPA 70B**).

Overcurrent conditions caused by ground-faults are usually more troublesome, in that they are normally not bolted faults, where there is an insulation void and the ungrounded conductor makes direct contact with the equipment grounding system, as if the ungrounded conductor were physically bolted to the equipment grounding system. It is much more likely that the insulation has absorbed contaminants, such as dirt combined with moisture, causing conducting or tracking paths through the conductor insulation. If this occurs near a grounded surface (metal raceway, metal enclosure, etc.), an arc may develop. And then the ground-fault current can return to the electrical system source, and, hopefully, clear the overcurrent device quickly to avoid damage. Unfortunately, this will probably not be the case. The arcing ground-fault just described will have a voltage-drop associated with it, and this will reduce the amount of ground-fault current produced in this circuit. Now, the overcurrent device may not see a sufficient amount of current to cause it to promptly clear. Add to this problem the fact that the arcing ground-fault is intermittent in nature, and the overcurrent device may not clear at all. This is the reason why the equipment grounding system, and the overcurrent protection must be carefully designed to limit the duration of arcing faults because they will be the most common faults in a distribution system.

It is a common misconception that when a megohmmeter is used for testing insulation, the test is to determine the quality of the insulation. When, in fact, the test is performed to determine the contaminants that have been absorbed by the insulation, typically, dirt, combined with moisture. This may lead to conducting paths, known as 'tracking paths'. It is these 'tracking paths' that typically lead to ground-faults, and even possibly, to short-circuits.

A good way to test insulation resistance is the 'Dielectric Absorption Ratio Test' and the 'Polarization Index Ratio Test'. These testing procedures are for rotating machinery (ANSI/ IEEE 43).

The Dielectric Absorption Ratio is a test where the insulation is stressed for a period of 60 seconds, and then, for 30 seconds. The 60-second test reading is divided by the 30-second test reading, and a ratio is established. A value of 1.4, or higher, is indicative of insulation that is clean and dry. The most common voltage used for this test is 500 to 1000 volts. Ratios of less than 1.1 are considered to be poor, ratios of 1.1 to 1.25 are questionable, while ratios of 1.25 to 1.4 are only considered to be fair.

The Polarization Index Ratio Test is where the insulation is stressed for 10 minutes, and then for 1 minute. A ratio of two to four would indicate that the condition of the insulation is good.

For low-voltage cable testing, the test is performed on each conductor with respect to ground, as well as to adjacent conductors. This test is done at 500 volts (direct current) for 300-volt rated cable and 1000 volts (direct current) for 600-volt rated cable. The cable is stressed for a period of one minute. The test results are compared with manufacturer's published data.

An important consideration in insulation resistance testing is the temperature conversion factors of ANSI/NETA Mts-2011.

Certainly, the ambient temperature will have an effect on the resistance readings.

For example, if the testing is done at the fallowing ambient temperatures, the test results are converted as follows.

Temperature		**Conversion Factor**
40°C,(104°F)	-	2.50
30°C,(86°F)	-	1.58
25°C,(77°F)	-	1.25
20°C,(68°F)	-	1.00
15°C,(59°F)	-	0.81
0°C,(32°F)	-	0.40

These conversion factors are for apparatus that has solid insulation.

Source: NFPA 70B-Recommended Practice for Electrical Equipment Maintenance (2016) Table 11.21. 3.1(b).

The following Sections relate to the terms associated with overcurrent or overload conditions.

110.9 – Interrupting Rating – This section involves the requirement that equipment that is intended to interrupt current under fault conditions must have an interrupting rating, at nominal circuit voltage, at least equivalent to the available fault-current at the line terminals of the equipment. So, the available fault-current must be calculated and applied accordingly.

If the equipment is intended to interrupt current at other than fault levels, such as the locked-rotor currents of motors, this equipment must have an interrupting rating, at nominal voltage, at least equivalent to the current to be interrupted (430.110(C)(1)), (Table 430.251(A)), (Table 430.251(B)).

110.10 – Circuit Impedance, Short-Circuit Current Ratings, and Other Characteristics–This section requires that the total circuit impedance be known, as well as the equipment short-circuit current ratings (including the equipment supply conductor insulation withstand ratings), and the overcurrent protective device characteristics, so that a fault may be properly cleared without the occurrence of extensive damage to the electrical equipment of the circuit.

In many cases, the equipment short-circuit current rating is identified through markings on the equipment, such as panelboards, motor control centers, busways, and surge-protective devices. Conductor insulation short-circuit withstand ratings are identified in a Table in this book.

110.14(C) – Temperature Limitations – The temperature rating that is associated with the ampacity of a conductor may not exceed the lowest temperature limit of any termination, device, or conductor.

This provision mirrors the information in the UL Green and White Books, which predate the NEC reference as it relates to terminations. For circuits rated up to 100 amperes, or identified for conductors 14 AWG through 1 AWG, the terminations may be used only for conductors rated for 60°C. (110.14(C)(1),(a),(1)). However, conductors with higher temperature ratings may be used, but only if their ampacity is based on the 60°C. ampacity of the conductor (110.14(C)(1),(a),(2)).

If the termination is listed and identified for a higher temperature, the conductor ampacity may be based on the higher temperature (110.14(C)(1),(a),(3)).

For motors with Design letters B, C, or D, conductors having an insulation rating of 75°C. may be used (110.14(C)(1),(a),(4)).

For conductors larger than 1 AWG, or, for circuits over 100 amperes, conductors with insulation temperature ratings of 75°C. shall be used (110.14(C)(1),(b),(1)).

Conductors with higher temperature ratings may be used, providing the conductor ampacity does not exceed the 75°C. ampacity of the conductor, unless the termination is listed and identified for a higher temperature (110.14(C)(1),(b),(2)).

110.24 – This Section, which is relatively new, requires that service equipment, in other than dwelling occupancies, be legibly marked in the field with the maximum available fault-current. The date that the fault-current calculation was made must be included on the field marking and the calculation must be documented and made available to the appropriate personnel authorized to design, install, inspect, maintain, or operate the system.

If modifications are made at a later date, the available fault current must be verified or recalculated to identify the new available short-circuit current and the field markings must be adjusted accordingly.

There is an exception that negates this requirement in industrial establishments, where conditions of maintenance and supervision ensure that qualified persons service the installation.

240.4- This section involves the overcurrent production of conductors in accordance with their ampacities as specified in 310.15.

An Informational Note identifies ICEA-P-32-382-2007. This information, from the Insulated Cable Engineers Association, identifies the allowable short-circuit currents for insulated copper and aluminum conductors.

Once again, the conductor overcurrent protection applies to overload, short-circuit, and ground-fault conditions.

So, this does not always mean that a conductor must be provided with overcurrent protection in accordance with the conductor ampacity. For example, where there is a 'Power Loss Hazard', and where the interruption of power may lead to additional hazards, the conductor overload protection may be omitted. But, the conductor must be provided with short-circuit protection. This means that the appropriate short-circuit calculations must be made and the conductor overcurrent protection must protect against these fault conditions (240.4(A)).

A very common condition applies where the conductor ampacity does not correspond to a standard overcurrent device rating as identified in 240.6(A), and the conductor is protected with the next standard size of overcurrent device. This is acceptable up to 800 amperes, and, where the protected conductors are not part of a branch-circuit that supplies receptacles for cord-and-plug connected portable loads (240.4(B)).

For small conductors, the overcurrent protection is limited to:

7 amperes for 18 AWG- copper
10 amperes for 16 AWG- copper
15 amperes for 14 AWG- copper
20 amperes for 12 AWG- copper
25 amperes for 10 AWG- aluminum or copper-clad aluminum
30 amperes for 10 AWG- copper

Where the load is continuous on these conductors, the load may not exceed 80% of the circuit rating.

Tap conductors are permitted to be protected against overcurrent when supplying household ranges and cooking appliances, as well as other loads. Electric ranges, wall-mounted electric ovens, and counter-mounted electric cooking units may be supplied by conductors having an ampacity of not less than 20 amperes, where they are tapped to a 50 ampere branch-circuit (210.19(A)(3), Exception No 1).

240.5(B)(2) – Fixture wire may be tapped to a branch-circuit conductor of a branch-circuit in accordance with the following.

1. 20 ampere circuits – 18 AWG up to 50 feet (15m)
2. 20 ampere circuits – 16 AWG up to 100 feet (30m)
3. 20 ampere circuits – 14 AWG and larger
4. 30 ampere circuits – 14 AWG and larger
5. 40 ampere circuits – 12 AWG and larger
6. 50 ampere circuits – 12 AWG and larger

240.21 – Generally requires that overcurrent protection for ungrounded conductors be provided where these conductors receive their supply.

However, this Section recognizes various tap rules for branch-circuit conductors, as well as feeder taps, as referenced in 210.19, 210.20, and 240.21(B)(1) through (6).

Section 368.17(B), Exception permits busway taps in industrial establishments. In this case, a busway of a lower ampacity may be tapped into a larger busway where the length is restricted to 50 feet (15m), and the smaller busway has an ampacity of at least one-third of the rating or setting of the overcurrent device upstream of the tap. Also, the smaller busway must not be in contact with combustible material.

368.17(C) – where busways are used as feeders and supply feeders or branch-circuits, the busway taps must contain the overcurrent devices required to protect the feeders or branch-circuits.

However, the various tap rules as permitted by 240.21 are Exceptions to 368.17(C), as well as Exceptions for cord connected luminaires, where the overcurrent device is part of the cord plug, and where luminaires are plugged into the busway, with the overcurrent device as part of the luminaire.

Another Exception recognizes a branch-circuit overcurrent device that is plugged into the busway and supplies a readily accessible disconnect. In this case, a floor operable disconnect is not required.

240.4 (E)(6) recognizes single motor taps from 430.53(D).

240.4(F)-involves the protection of transformer secondary conductors for single-phase transformers with 2-wire single-voltage secondaries, and three-phase, Delta-to-Delta connected transformers with 3-wire single-voltage secondaries. In these cases, the secondary conductors are permitted to be protected by the primary overcurrent device, provided this protection has been sized in accordance with 450.3, and does not exceed the value by multiplying the secondary conductor ampacity by the secondary-to-primary transformer voltage ratio.

For example, a 25 kVA, single-phase, 480 volt primary, and a 120 volt secondary

$$\frac{25,000\,VA}{480\,Volts} = 52\,amperes$$

Full-load primary current – 52 amperes

$$52\ amperes$$
$$\underline{\times 1.25\ (450.3(B))}$$
$$65\ \text{amperes, the next standard size} = 70\ amperes$$

$$\frac{25,000\,VA}{120\,Volts} = 208\,amperes$$

Full-load secondary current – 208 amperes
Secondary-to-primary transformer voltage ratio

$$\frac{120\,Volts}{480\,Volts} = \frac{1}{4}\,or\ .25$$

$$\frac{70}{.25} = 280\ amperes$$

300 kcmil copper = 285 amperes @ 75°C.

(240.5) This Section involves the overcurrent protection of flexible cords, flexible cables, and fixture wires.

The general rule is that a flexible cord or cable is to be protected with an overcurrent device in accordance with their ampacity as referenced in Table 400.5(A)(1) and Table 400.5(A)(2).

For fixture wire, the overcurrent protection is in accordance with Table 402.5.

This overcurrent protection may be provided as supplementary overcurrent protection in accordance with 240.10 (240.5(A)).

If a listed appliance or luminaire is provided with a supply cord, the overcurrent protection may be in accordance with the equipment listing instructions (240.5(B)(1)).

240.21 – This Section addresses the various tap rules where an overcurrent device is not provided where a conductor receives its supply.

Effective on January 1, 2020, in order to accomplish arc energy reduction for fuses rated 1200 amperes or higher, the fuse clearing time must not exceed 0.07 seconds (4.19 cycles at 60Hz) at the available arcing current (240.67(B)), or the methods of reducing the clearing time of overcurrent devices as identified in 240.87(B).

240.87 – This Section involves 'Arc Energy Reduction' where the highest continuous-current trip-setting of the overcurrent device installed in a circuit breaker is rated or can be adjusted is 1200 amperes or higher.

The concept of reducing arc energy and the means of accomplishing this condition is critical in mitigating an arc-flush hazard by reducing the duration of the arcing fault.

Methods for reducing the clearing time of the overcurrent device are indentified in 240.87(B).

1. Zone-selective interlocking
2. Differential relaying
3. Energy-reducing maintenance switching with local status indicator
4. Energy-reducing active arc-flash mitigation system
5. An instantaneous trip-setting that is less than the available arcing (fault) current
6. An instantaneous override that is less than the available arcing (fault) current
7. An approved equivalent means

The clearing time of the overcurrent protection is an extremely important consideration in reducing an arc-flash hazard while a worker is within the arc-flash boundary.

For example, 240.87(B)(3) recognizes this through the use of energy reducing maintenance switching, where the worker may set the circuit breaker tripping mechanism so that there is no time-delay in order to reduce the clearing time while within the arc-flash boundary, as defined in NFPA-70 E-2015.

In my opinion, the most effective method of reducing the clearing time of a circuit breaker for the purpose of Arc Energy Reduction to comply with 240.87(B), is Provision 4 of this Section. This is an energy-reducing active arc flash mitigation system.

This type of protection permits a technician to set an energy-reducing maintenance switch in order to reduce the clearing time of the circuit breaker while performing maintenance with the person working within the arc-flash boundary of NFPA70E. A power system analysis must be performed to determine the fault current that will flow through the circuit breaker associated with the Arc-flash Reduction Maintenance System unit.

Transient load current must be determined, including motor inrush current and transformers inrush current.

The fault current will normally be arcing and not bolted fault current. A pickup setting for the Arcflash Reduction Maintenance System will be below 75% of the calculated arcing current and above the total transient load current. A table in IEEE STD 1584TM-2002 and formulas are used to calculate arcing fault current.

250.122–This section involves the sizing of Equipment Grounding Conductors. The various types of equipment grounding conductors are referenced in 250.118.

Wire-type equipment grounding conductor sizes are required to be in accordance with the minimum size listed in Table 250.122. They are never required to be larger than the ungrounded or grounded circuit conductors (250.122(A)).

However, if the ungrounded conductors are increased in size to compensate for voltage-drop, or the effects of adjacent load carrying conductors, there must be a proportional increase in the size of the equipment grounding conductor (250.122(B)).

Example

Circuit size – 200 amperes
Ungrounded conductor size – 3/0 AWG copper-However, due to voltage-drop, conductors are increased in size to 4/0 AWG copper
Equipment grounding conductor – 6 AWG copper-Table 250.122

$$\frac{4/0 - 211,600\,circular\,mils}{3/0 - 167,800\,circular\,mils} = 1.26$$

$$6\,AWG - copper - 26,240 \;\; circular\,mils$$
$$\underline{\times 1.26}$$
$$33,062 \;\; circular\,mils$$

33,062 circular mils – 4 AWG – 41,740 circular mils, Equipment Grounding Conductor - 4 AWG copper (Table 8-Chapter 9). However, as we have stated previously, the AHJ may permit the size of the EGC to be determined by a qualified person instead of a proportional increase in the EGC size.

Where multiple circuits are installed in the same raceway, cable, or cable tray, the equipment grounding conductor size is based on the rating of the largest overcurrent device protecting the multiple circuits (250.122(C)).

For motor circuits, the equipment grounding conductor size is based on the rating of the overcurrent device protecting the motor circuit (250.122(D)(1)).

If the motor overcurrent device is an instantaneous-trip circuit breaker or a motor short-circuit protector, the equipment grounding conductor size from 250.122(A) may be based on the equivalent size of a dual-element time-delay fuse (250.122(D)(2)).

For paralleled conductors (310.10(H)) installed within the same raceway or cable tray, a single equipment grounding conductor, sized in accordance with 250.122, based on the size of the circuit overcurrent device, is acceptable.

If the paralleled conductors are installed in multiple raceways, an equipment grounding conductor must be installed in each raceway. The size of the equipment grounding conductor in each raceway is based on the rating of the overcurrent device in accordance with 250.122.

Of course, if the paralleled conductors are installed in metal raceways, auxiliary gutters, or metal cable trays in accordance with 250.118, and 392.60(B), the metal raceways, auxiliary gutters, or metal cable trays may serve as the equipment grounding conductor (250.122(F)(1)).

If multiconductor cables are installed in parallel, the equipment grounding conductor(s) in each cable shall be connected in parallel.

If multiconductor cables are installed in parallel in the same raceway, auxiliary gutter, or cable tray, a single equipment grounding conductor that is sized in accordance with 250.122, shall be permitted in combination with the equipment grounding conductors provided within the multiconductor cables. And the equipment grounding conductors will be connected together.

310.15 – Ampacities for Conductors Rated 0-2000 volts – 310.15(A)(1) recognizes that conductor ampacities may be determined by the Tables in 310.15(B), or under engineering supervision (310.15(C)).

Calculating conductor ampacities under engineering supervision may result in the use of smaller conductors for a given load than using the Tables in 310.15(B). A similar condition applies when determining the maximum current for Photovoltaic Source Circuit Current, where the generating capacity is 100 kW or greater (690.8(A)(1)). The use of this method will typically result in reduced installation costs and the calculations must be done by a licensed professional electrical engineer.

The most commonly used Table of conductor ampacities is Table 310.15(B)(16), and the basis for the expressed ampacities are conductor temperature ratings of 60°C., 75°C., and 90°C. The ambient temperature is identified as 30°C. (86°F), and the ampacities are based on having no more than 3 current carrying conductors in a raceway, cable, or directly buried.

For ambient temperatures above 30°C (86°F), Table 310.15(B)(2),(a) specifies the appropriate correction factors that must be applied to the normal conductor ampacity in order to protect the conductor from overheating and insulation damage. Conductors have a positive temperature coefficient, that is, as the ambient temperature increases, the conductor resistance also increases. Table 310.15(B)(2),(b) has ambient temperature correction factors based on 40°C. (104°F) temperatures.

Proximity effects, or the effects of adjacent current-carrying conductors within a raceway, cable, or directly buried where their number exceeds 3, is addressed in Table 310.15(B)(3),(a). Further information relating to this adjustment factor is covered in Annex B, where there are more than 3 current-carrying conductors installed in a raceway or cable with load diversity.

It should be noted that this adjustment factor applies where the conductors are installed together without maintaining spacing for lengths exceeding 24 inches (600mm).

Neutral conductors that carry only the unbalanced current from the ungrounded conductors of the same circuit are not considered current-carrying conductors for the application of this adjustment factor.

However, where a 3-wire circuit is derived from a three-phase, 4-wire, Wye connected system, consisting of 2 phases and a neutral conductor, the neutral conductor will always carry current, and it is considered to be a current-carrying conductor (310.15(B)(5)).

Also, on a three-phase, 4-wire, Wye connected system where the major portion of the load is nonlinear, the neutral is considered to be a current-carrying conductor (310.15(B)(5)).

Cable trays are not raceways by definition, so the application of conductor ampacities is in accordance with 392.80.

Grounding or bonding conductors are not considered to be current-carrying conductors (310.15(B)(6)).

430.22- The Section involves the conductors that supply a continuous-duty motor. The Note of Table 430.22(E) states that any motor application shall be considered as continuous-duty, unless the nature of the apparatus it drives is such that the motor will not operate continuously with load, under any condition of use. Continuous-Duty is defined in Article 100 as 'operation at a substantially constant load for an indefinitely long time'. So, conductors that supply a continuous-duty motor, in accordance with this definition, must have an ampacity of at least 125% of the motor full-load current. And, the motor full-load current is in accordance with the appropriate Table from Article 430, and not from the motor nameplate (Table 430.247, 248, 249, 250), (430.6(A)(1)).

In this regard, the motor circuit conductors will have an ampacity that is approximately the same as the motor overload protective device(s) (430.32).

430.28 – This Section applies to motor feeder taps. These tap rules are similar to the feeder tap rules of 240.21(B)(1),(2).

1. The tap conductors may be not more than 10 feet (3m) in length, be protected from physical damage by an enclosed controller or a raceway, and, where installed in the field, the tap conductors must be protected on their line side by an overcurrent device that has a rating or setting of no more than 1000% of the tap conductor ampacity,
2. The tap conductors are not more than 25 feet (75m) in length, they are protected from physical damage, and they have an ampacity of at least one-third of the ampacity of the feeder conductors.
3. The tap conductors have an ampacity not less than that of the feeder conductors.

430.110 – This Section involves the ampere rating of a motor disconnecting means. 430.110(A) specifies that the ampere rating must be at least 115% of the full-load current rating of the motor.

The horsepower rating must be at least equal to the locked-rotor current of the motor in accordance with the locked-rotor currents expressed in Table 430.251(A) (single-phase) or Table 430.251(B) (polyphase-Design B, C, or D), (430.110(C)(1)).

There is a connection between 430.110(C)(1) and 110.9, as it relates to the interrupting rating of equipment at other than fault levels, such as the locked-rotor currents of motors.

440.6 – This Section applies to sizing the conductors for a hermetic refrigerant motor-compressor. Generally, the conductor ampacity is based on the rated-load current that is marked on the equipment. The marked rated-load current is also used to determine the ampere rating of the disconnecting means, motor controller, the overcurrent protection, and the separate motor overload protection (440.6(A)).

For multimotor equipment, the full-load current is, once again, determined by the full-load current marked on the equipment nameplate (440.6(B)).

440.12–This Section applies to the ampere rating of the disconnecting means for a hermetic refrigerant motor-compressor. If the 'branch-circuit selection current' is identified on the equipment nameplate, it must be used in determining the rating of the circuit components if this current is greater than the nameplate rated-load current (440.12(A)).

The ampere rating of the disconnecting means must be at least 115% of the greater of these two current values.

To determine the horsepower rating of the disconnecting means in order to satisfy the requirements of 110.9 and 430.109(A)(1), the horsepower rating is selected from Tables 430.248 (single-phase), 430.249 (two-phase), or 430.250 (three-phase), in accordance with the greater of the branch-circuit selection current or the rated-load current. Also, the horsepower rating from Tables 430.251(A) or 430.251(B) is identified, corresponding to the locked-rotor current. If the branch-circuit selection current or the nameplate rated-load do not correspond to the currents identified in these Tables, the next higher horsepower rating must be selected (440.12(A)(2)).

445.12 – Constant-Voltage generators are protected from overcurrent by inherent design, circuit breakers, fuses, protective relays, or other identified overcurrent protective means suitable for the conditions of use.

445.13–The conductors from the generator output terminals to the first device containing overcurrent protection must have an ampacity of 115% of the nameplate current rating of the generator. Of course, where the generator is supplied with identified overcurrent protection, this does not apply.

And, where the generator has a listed overcurrent protective device, or a current transformer with on overcurrent relay, conductors may be tapped from the load side of the overcurrent protection as referenced in 240.21(B).

445.20–This Section applies to GFCI protection for receptacles on 15 kW or smaller portable generators.

The GFCI protection must be listed for personnel that is integral to the generator or receptacle.

If the generator neutral is not bonded to the generator frame, listed GFCI protection must be provided for 125 volt and 125/250 volt receptacle outlets (445.20(A)).

If the generator neutral is bonded to the generator frame, GFCI protection is required for all 125 volt, 15 and 20 ampere receptacle outlets (445.20(B)).

Where the portable generator is used for temporary installations, these requirements are addressed in 590.6(A)(3). There is an exception for 15 kW or smaller generators that were manufactured or remanufactured before January 1, 2011, where listed cord sets or devices that have listed ground-fault circuit interrupter protection for personnel may be used in lieu of the devices that are integral with the generator.

Section 250.34(A),(B) recognizes that the frame of a portable or vehicle-mounted generator need not be connected to a grounding electrode, where the generator supplies equipment mounted on the generator, and/or cord-and-plug connected equipment through receptacles mounted on the generator, and the non-current-carrying metal parts of equipment and the equipment grounding terminals of receptacles are bonded to the generator frame. A grounding electrode at the generator may increase possibility of a shock hazard or electrocution to the operator of a portable device plugged into the generator outlet where a fault in the portable device allows the current to flow through the operator, the earth, and through the grounding electrode to the generator winding.

For vehicle-mounted generators, the frame of the generator must be bonded to the vehicle frame.

450.3–This Section involves the overcurrent protection of transformers at voltages of over 1000 volts and 1000 volts or less. And, this is an example of 90.1(A), which states that 'this Code is not intended as a design specification or an instruction manual for untrained persons'.

The overcurrent protection for a transformer may be affected by using a protective device rated or set at the protection levels expressed in 450.3(A) or 450.3(B).

While it is true that the normal magnetizing inrush currents for power transformers may reach 12 times the transformer full-load current for 6 cycles (0.096 seconds), and possibly, as many of 25 times for up to 0.01 seconds, the need to provide the proper overcurrent protection is based on the careful analysis of the operating characteristics of the selected overcurrent device in relationship to the expected inrush current, and not merely relying on the percentages in NEC Tables 450.3(A) and 450.3(B).

The use of time-delay fuses for primary and secondary overcurrent protection, at possibly 125% of the primary and secondary full-load current rating, may provide the required overcurrent protection. Dual-element time-duly fuses are designed to carry five times their rating for ten seconds.

Properly sized inverse-time circuit breakers at 125% of the primary and secondary full-load currents may provide the desired overcurrent protection (450.3(B)).

Another important consideration applies to transformers that are connected Delta-to-Wye.

There is a 30 degree phase-shift from primary to secondary. That is, the primary voltage will lead the secondary voltage by 30 degrees. Overloads, short-circuits, or ground-faults on the secondary side of the transformer may not clear the primary overcurrent device. For this reason, a properly sized secondary overcurrent device, in accordance with overload and fault conditions, is a must, even if Tables 450.3(A) and (B) do not always require secondary overcurrent protection.

450.3(A) applies to the overcurrent protection of transformers over 1000 volts.

The primary protection is based on the transformer rated impedance, as well as whether this protection is in the form of a circuit breaker or a fuse. And if the transformer is installed in a supervised location or in any location.

Where the transformer rated impedance is not more than 6% in any location, a circuit breaker may be set at 600% of the transformer primary full-load current rating. A fuse may have a rating of 300%.

If the transformer rated impedance is more than 6%, but not more than 10%, the circuit breaker may be set at 400%, and the fuse rating may be 300%.

Note 1 of Table 450.3(A) permits the setting or rating of the circuit breaker or fuse to be the next higher commercially available setting or rating where the required setting or rating does not correspond to a standard setting or rating.

For supervised locations, and any transformer rated impedance, the circuit breaker setting may be 300% of the transformer primary full-load current rating and the fuse rating may be 250%. Note 1 of Table 450.3(A) applies here as well.

In a supervised location, and a transformer rated impedance of not more than 6%, the circuit breaker setting may be 600% of the transformer primary full-load current rating and a fuse may be rated at 300%.

If the transformer rated impedance is more than 6% and not more than 10%, the circuit breaker setting may be 400% of the transformer primary full-load current rating, a fuse may be rated at 300%.

The secondary protection for voltages of over 1000 volts in <u>any</u> location, where the transformer rated impedance is not more than 6%, may be 300% for a circuit breaker setting, or 250% for a fuse. Note 1 of Table 450.3(A) applies here.

Where the transformer rated impedance is more than 6% and not more than 10%, the circuit breaker setting may be 250%, and the fuse rating may be 225%. Note 1 of Table 450.3(A) applies here.

In a supervised location, and <u>any</u> transformer rated impedance, secondary overcurrent protection is not required.

Note 5 of Table 450.3(A) permits the secondary overcurrent protection to be omitted if the transformer is equipped with coordinated thermal overload protection by the manufacturer.

For transformer secondary voltages of 1000 volts or less, the circuit breaker or fuse rating may be 125% of the transformer secondary full-load current rating, where the transformer is installed in <u>any</u> location. Note 1 of Table 450.3(A) applies here.

In a <u>supervised</u> location, and any transformer rated impedance, secondary overcurrent protection is not required.

In a <u>supervised</u> location, where the transformer rated impedance is not more than 6%, the circuit breaker or fuse rating may be 250% of the transformer full-load secondary current rating. Note 5 applies here.

In a <u>supervised</u> location, where the transformer rated impedance is more than 6% and not more than 10%, the circuit breaker or fuse rating may be 250% of the transformer full-load secondary current rating. Note 5 applies here.

For medium voltages, a general purpose current-limiting fuse is capable of interrupting all currents from the rated interrupting current down to the current that causes the melting of the fusible element in one hour. These fuses have either an 'E' or an 'X' rating.

'E' rated fuses 100E and below have a fuse element that must melt in 300 seconds at 200% to 240% of the fuse rating (ANSI C37.46).

'E' rated fuses above 100E have a fuse element that must melt in 600 seconds at 220% to 264% of the fuse rating (ANSI C37.46).

Table 450.3(B) applies to the overcurrent protection of transformers at 1000 volts, and less. This Table is divided into 3 current categories.

Current Categories

Primary-only Protection	*-currents of 9 amperes or more*	*currents less than 9 amperes*	*currents less than 2 amperes*
	125%	167%	300%

Where the application of these percentages results in an ampere rating that does not correspond to a standard size of overcurrent device (240.6(A)), the next higher standard rating is permitted.

In this case, no secondary overcurrent device is required.

Primary and secondary protection	*currents of 9 amperes or more*	*currents less than 9 amperes*	*currents less than 2 amperes*
Primary	250%	250%	250%
Secondary	125%	167%	–

For secondary currents of 9 amperes, or more, where 125% of the full-load secondary current does not correspond to a standard overcurrent device rating (240.6(A), the next standard rating may be used.

The following sections have been selected for review because they relate to the interrupting ratings of fuses and circuit breakers, circuit impedance, and short-circuit ratings of conductors and equipment in order to permit the circuit protective devices to clear short-circuits or ground-faults without causing extensive damage to the electrical equipment. Available fault-current identification at service equipment, overcurrent protection of conductors, various conductor tap rules, where the conductor overload protection is not provided, but short-circuit and ground-fault protection for the tap conductors is not compromised, is also a part of these sections.

Remember, the term 'overcurrent' is defined in Article 100 as 'any current in excess of the rated current of equipment or the ampacity of a conductor.' This condition may be caused by an overload, short-circuit, or ground-fault.

In order to comply with the concept of 'overcurrent protection', conductor and equipment withstand ratings must be determined.

Also, interrupting ratings of overcurrent devices, motor disconnecting switches and controllers must be capable of interrupting the appropriate fault-current or locked-rotor current in order to avoid damage to these devices, or to the connected conductors and equipment.

The following Table contains information relating to maximum withstand ratings of conductor insulation, based on the use of 75° C. rated insulation, and a maximum insulation temperature of 150° C.

In addition, a second chart identifies the amount of current that would produce a maximum temperature of 250° C. in a terminal device for different time constants, ranging from 5 seconds down to .002 second (1/8 cycle). This relates to the rapid heating and cooling of the terminal, causing expansion and contraction within the terminal, without affecting the terminal.

And, finally, the amount of current that would cause a conductor to fuse or melt as it reaches a temperature of 1083° C.

The time constants in the Tables are based on the normal continuous current ratings of the conductors, in sizes ranging from #14 AWG through 500 kcmil at 75° C. And the various short-time current ratings at 5 seconds, 1 second, 1 cycle (.016 seconds), 1/2 cycle (.008 seconds), 1/4 cycle (.004 seconds), and 1/8 cycle (.002 seconds). As we have stated, this information is useful in determining whether a conductor and equipment has proper overcurrent protection, that is, protection against overloads, short-circuits, or ground-faults to satisfy the requirements of NEC Sections 240.4 and 110.10.

Also, we may use this information to determine the conductor termination integrity, based on a 250°C. temperature limit. And, lastly, we may determine the validity of an equipment grounding conductor, based on the level of current and time that it would take to reach the conductor melting temperature (1083 ° C). We would have to determine the amount of ground-fault current that the equipment grounding conductor is expected to carry until the ground-fault is cleared by the circuit overcurrent device. Remember, that NEC Table 250.122 only identifies the minimum equipment grounding conductor size.

The following examples will identify a means of calculating the 150° C, 250° C, and 1083° C. conductor temperature of a No. 6 AWG - 13.30 mm^2 copper conductor. Please refer to the Tables to determine this information for other conductor sizes and different time constants.

This information may be used to determine whether proper overcurrent protection (time and current) has been provided for the purpose of protecting conductor insulation, as well as determining the integrity of conductor terminations. And, to determine the validity of the equipment grounding system.

Insulation Withstand Ratings 150° C. Maximum

AWG	Normal 75°C.	5 Second 150°C.	1 Second 150°C.	1 cycle – .016 sec. 150°C.	1/2 cycle – .008 sec. 150°C.	1/4 cycle – .004 sec. 150°C.	1/8 cycle – .002 sec. 150°C.
14	20A	97A	217A	1,715A	2,425A	3,429A	4,850A
12	25A	155A	347A	2,740A	3,875A	5,480A	7,750A
10	35A	246A	550A	4,349A	6,150 A	8,697A	12,300A
8	50A	397A	888A	6,912A	9,775A	13,824A	19,550A
6	65A	621A	1,389A	10,978A	15,525A	21,956A	31,050A
4	85A	988A	2,209A	17,466A	24,700A	34,931A	49,400A
3	100A	1,245A	2,784A	22,008A	31,125A	44,017A	63,450A
2	115A	1,571A	3,513A	27,772A	39,275A	55,543A	78,550A
1	130A	1,981A	4,430A	35,019A	49,525A	70,039A	99,050A
1/0	150A	2,499A	5,588A	44,176A	62,475A	88,353A	124,950A
2/0	175A	3,150A	7,044A	55,685A	78,750A	111,369A	157,500A
3/0	200A	3,972A	8,882A	70,216A	99,300A	140,431A	198,600A
4/0	230A	5,008A	11,198A	88,530A	125,200A	177,060A	250,400A
250 kcmil	25SA	5,917A	13,231A	104,599A	147,925A	209,198A	295,850A
300 kcmil	285A	7,101A	15,878A	125,529A	177,525A	251,058A	355,050A
350 kcmil	310A	8,284A	18,524A	146,442A	207,100A	292,884A	414,200A
400 kcmil	335A	9,467A	21,169A	167,354A	236,675A	334,709A	473,350A
500 kcmil	380A	11,834A	26,462A	209,198A	295,850A	418,395A	591,700A

Example

Insulation Withstand Rating

No. 6 AWG copper – 65 amperes (continuous) – 75°C

No. 6 AWG copper – 26,240 circular mils – 13.30mm^2

I^{2}T – ampere – squared seconds

I^{2}T – one ampere for every 42.25 circular mils of conductor cross- sectional area for 5 seconds

$$No.6 \ AWG - \frac{26,240 \, circular \, mils}{42.25} = 621 \, amperes - 5 \, seconds$$

To determine the insulation withstand rating for 1 cycle (.016 seconds), the calculation is as follows:

621 amperes × 621 amperes x 5 seconds = 1,928,205

$$\frac{1,928,205}{.016} = 120,512,812$$

$$\sqrt{120,512,812} = 10,978 \, amperes$$

Therefore, the one-cycle insulation withstand rating is 10,978 amperes, which will produce a temperature of 150°C. This is the maximum temperature that the insulation can safely withstand without damage.

Terminal Withstand Ratings 250° C. Maximum

AWG	Normal 75°C.	5 Second 250°C.	1 Second 250°C.	1 cycle – .016 sec. 250°C.	1/2 cycle – .008 sec. 250°C.	1/4 cycle – .004 sec. 250°	1/8 cycle – .002 sec. 250°C.
14	20A	112A	250A	1,980A	2,800A	3,960A	5,600A
12	25A	178A	398A	3,147A	4,450A	6,293A	8,900A
10	35A	282A	631A	4,985A	7,050A	9,970A	14,100A
S	50A	449A	1,004A	7,937A	11,225A	15,875A	22,450A
6	65A	714A	1,597A	12,622A	17,850A	25,244A	35,700A
4	85A	1,136A	2,540A	20,082A	28,400A	40,164A	56,800A
3	100A	1,459A	3,262A	25,792A	36,475A	51,583A	72,950A
2	115A	1,806A	4,038A	31,926A	45,150A	63,852A	90,300A
1	130A	2,277A	5,092A	40,252A	56,925A	80,504A	113,850A
1/0	150A	2,873A	6,424A	50,788A	71,825A	101,576A	143,650A
2/0	175A	3,622A	8,099A	64,029A	90,550A	128,057A	181,100A
3/0	200A	4,566A	10,210A	80,716A	114,150A	161,432A	228,300A
4/0	230A	5,758A	12,875A	101,788A	143,950A	203,576A	287,900A
250 kcmil	255A	6,803A	15,212A	120,261A	170,075A	240,522A	340,150A
300 kcmil	285A	8,163A	18,253A	144,303A	204,075A	288,606A	408,150A
350 kcmil	310A	9,524A	21,296A	168,362A	238,100A	336,724A	476,200A
400 kcmil	335A	10,884A	24,337A	192,404A	272,100A	384,808A	544,200A
500 kcmil	380A	13,605A	30,422A	240,505A	340,125A	481,009A	680,250A

<u>Example</u>

<u>Terminal Withstand Rating</u>

No. 6 AWG copper – 65 amperes (continuous) – 75°C
No. 6 AWG copper – 26,240 circular mils – 13.30mm²
I^2T – one ampere for every 36.75 circular mils of conductor cross-sectional area for 5 seconds

$$No.6\ AWG - \frac{26,240\,circular\,mils}{36.75} = 714\,amperes - 5\,seconds$$

To determine the conductor terminal withstand rating for 1 cycle (.016 seconds), the calculation is as follows:

714 amperes × 714 amperes × 5 seconds = 2,548,980

$$\frac{2,548,980}{.016} = 159,311,250$$

$$\sqrt{159,311,250} = 12,622\,amperes$$

Therefore, the one-cycle (.016 seconds) terminal withstand rating is 12,622 amperes, which will produce a temperature of 250°C. This is the maximum temperature that the terminal can safely withstand.

Fusing or Melting Current 1083°C. Maximum

AWG	Normal 75°C.	5 Second 1083°C.	1 Second 1083°C.	1 cycle – .016 sec. 1083°C.	1/2 cycle – .008 sec. 1083°C.	1/4 cycle – .004 sec. 1083°C.	1/8 cycle – .002 sec. 1083°C.
14	20A	254A	568A	4,490A	6,350A	8,980A	12,700A
12	25A	403A	901A	7,124A	10,075A	14,248A	20,150A
10	35A	641A	1,433A	11,331A	I6,025A	22,663A	32,050A
8	50A	1,020A	2,281A	18,031A	25,500A	36,062A	51,000A
6	65A	1,621A	3,625A	28,656A	40,525A	57,311A	81,050A
4	85A	2,578A	5,765A	45,573A	64,450A	91,146A	128,900A
3	100A	3,312A	7,406A	58,548A	82,800A	117,097A	165,000A
2	II5A	4,101 A	9,170A	72,461A	102,475A	144,922A	204,950A
1	130A	5,169A	11,558A	91,376A	129,225A	183,105A	258,950A
1/0	150A	6,523A	14,586A	115,311A	163,075A	230,623A	326,150A
2/0	175A	8,221A	18,383A	145,328A	205,525A	290,656A	411,050A
3/0	200A	I0,364A	23,175A	183,211A	259,100A	366,423A	518,200A
4/0	230A	13,070A	29,225A	231,047A	326,750A	462,094A	653,500A
250 kcmil	255A	15,442A	34,529A	272,979A	386,050A	545,957A	772,100A
300 kcmil	285A	18,530A	41,434A	327,567A	463,250A	655,134A	925,500A
350 kcmil	310A	21,618A	48,339A	382,156A	540,450A	764,312A	1,080,900A
400 kcmil	335A	24,707A	55,247A	436,762A	617,675A	873,524A	1,235,350A
500 kcmil	380A	30,883A	69,056A	545,939A	772,075A	1,091,879A	1,544,150A

Example

Fusing or Melting Current

No. 6 AWG copper – 65 amperes (continuous) – 75°C
No. 6 AWG copper – 26,240 circular mils – 13.30mm^2
I^2T – one ampere for every 16.19 circular mils of conductor cross-sectional area for 5 seconds
I^2T –ampere –squared seconds

$$No.6\ AWG - \frac{26,240\,circular\,mils}{16.19} = 1621\,amperes - 5\,seconds$$

To calculate the fusing or melting current for 1 cycle (.016 seconds), the calculation is as follows:

1621 amperes × 1621 amperes × 5 seconds = 13,138,205

$$\frac{13,138,205}{.016} = 821,137,813$$

$$\sqrt{821,137,813} = 28,656\,amperes$$

AWG - To Metric Conversion Chart

AWG	Circular Mil Area	Metric Size –MM^2
14	4,110 cm	2.08
12	6,530 cm	3.31
10	10,380 cm	5.261
8	16,510 cm	8.367
6	26,240 cm	13.30
4	41,740 cm	21.15
3	53,620 cm	26.67
2	66,360 cm	33.62
1	83,690 cm	42.41
1/0	105,600 cm	53.49
2/0	133,100 cm	67.43
3/0	167,800 cm	85.01
4/0	211,600 cm	107.20
250 kcmil	250,000 cm	126.68
300 kcmil	300,000 cm	152.01
350 kcmil	350,000 cm	177.35
400 kcmil	400,000 cm	202.68
500 kcmil	500,000 cm	253.35

$$e.g., \ 2AWG = \frac{66,360 \, circular \, mils}{1973.53} = 33.625mm^2$$

Therefore, the one-cycle fusing or melting current of this conductor is 28,656 amperes, which will produce a temperature of 1083°C. Based on these conditions, the conductor will fuse or melt in 1 cycle.

Example

Insulation Withstand Rating

150 °C.

WIRE SIZE - No. 2 AWG – 33.62 mm^2 – Copper
75 ° C. ampere rating is 115 amperes
No. 2 AWG = 66,360 circular mils

$$\frac{66,360 \, circular \, mils}{42.25} = 1,571 \ amperes$$

NOTE: I^{2}T, or ampere-squared seconds, equals one ampere for every 42.25 circular mils of the conductor cross-sectional area for 5 seconds.

The conductor insulation, based on a normal 75° C. operating temperature, will safely withstand a current flow of 1,571 amperes for up to 5 seconds with no insulation damage.

Now, let's examine a condition where this No. 2 AWG copper conductor is connected to a 100 ampere molded-case circuit breaker, which has a 1.5 cycle (0.025 seconds) clearing time under short-circuit conditions.

The question now becomes whether the conductor insulation can safely withstand the short-circuit current which is available at the line terminals of the circuit breaker.

Let's <u>assume</u> that a fault-current analysis determines that this available short-circuit current is 21,400 amperes.

Will this conductor insulation withstand this level of current for the 1.5 cycle period that it would take this circuit breaker to operate and safely clear this short-circuit current?

2 AWG, copper = 66,360 circular mils

$$\frac{66,360 \text{ cm}}{42.25} = 1,571 \text{ amperes for 5 seconds}$$

Then, we establish $I^2 T$, or ampere-squared seconds.
1571 amperes × 1571 amperes × 5 seconds = 12,340,205 ampere-squared seconds

Then:

$$\frac{12,340,205}{.025 \text{ seconds}} = 493,608,200$$
$$= 1.5 \text{ cycles}$$

Then:
We calculate the square root of 493,608,200 = 22,217 amperes (.025 seconds). So, the 1.5 cycle (.025 seconds) insulation withstand rating for our 2 AWG copper wire is 22,217 amperes. To calculate insulation withstand ratings for different time constants, just divide the ampere-squared seconds rating of the wire by the time constant relating to the clearing time of the overcurrent device, and then calculate the square root as above.

Of course, in this example, the available short-circuit current at the line terminals of the circuit breaker is, hypothetically, 21,400 amperes. This would establish the required interrupting rating of the 100 ampere molded-case circuit breaker to satisfy the provisions of Section 110.9. And if the insulation withstand rating for 1.5 cycles of the 2 AWG copper conductor is 22,217 amperes, we would have satisfied the conductor insulation withstand rating requirements of Sections 110.10 and 240.4.

A further fault-current analysis at any downstream equipment would determine the required withstand rating there, as well.

We can use this method to calculate the <u>terminal integrity</u> of disconnect switches, motor controllers, circuit breakers, etc., based on their normal

75°C. operating temperature and their maximum temperature limit of 250°C. For our 2 AWG copper conductor, the calculation is as follows:

2 AWG - 66,360 cm

Terminal integrity is based on 1 ampere for every 36.75 circular mils of the conductor cross-sectional area for 5 seconds, or

$\dfrac{66,360}{36.75} = 1,806$ amperes for 5 seconds

Now, how about 1.5 cycle (.025 seconds) as before when we calculated the 1.5 cycle insulation withstand rating for the 2 AWG copper wire, the calculation to establish terminal integrity is as follows:

1806 amps × 1806 amps × 5 seconds = 16,308,180
$\dfrac{16,308,180 \text{ ampere-squared seconds}}{.025 \text{ seconds (1.5 cycle)}} = 652,327,200$

Then:
We calculate the square root of 652,327,200, and that equals 25,541 amperes. So, we have deduced that if this terminal was subjected to a current of 25,541 amperes for 1.5 cycles, the terminal integrity would not be affected because the available short-circuit current at the circuit breaker terminals is 21,400 amperes, in this example. The concept of terminal integrity is based on the heating and cooling of the terminal device when subjected to a high level of fault current. The effects of this rapid heating and cooling will produce expansion and contraction in the terminal, which may destroy its integrity.

As far as calculating the fusing or melting current of conductors, which is useful in determining the proper sizes of grounding electrode conductors, equipment grounding conductors, and bonding conductors, Table 250.122 provides information on the minimum sizes of equipment grounding conductors, based on the rating or setting of the circuit overcurrent device. But it may be necessary to increase these minimum sizes because of circuit conditions. These conditions may include excessive circuit length (over 100 feet), or higher levels of ground-fault current. Therefore, sometimes the equipment grounding conductor size may be relative to its fusing or melting current. How can we calculate these fusing or melting currents? As in the previous examples of insulation and terminal integrity, we start by calculating the fusing current of the wire for 5 seconds. Using the same 2 AWG copper wire, the basis is: one ampere for every 16.19 circular mils of cross-sectional area for 5 seconds.

So:

No. 2 AWG – $\frac{66,360 \text{ cm}}{16.19}$ = 4,099 amperes for 5 seconds.

Once again, how about 1.5 cycle (.025 seconds)?
For 2 AWG copper wire, the calculation is as follows:

4,099 amps × 4,009 amps × 5 seconds = 84,009,005 ampere-squared seconds

$\frac{84,009,005}{.025 \text{ seconds (1.5 cycles)}}$ = 3,360,360,200

Then:
We calculate the square root of 3,360,360,200, and that equals 57,969 amperes.

So, we have deduced that at 57,969 amperes of load, the 2 AWG copper wire will begin to fuse or melt in 1.5 cycles (.025 seconds).

One way to use the information relating to fusing current, and the relationship to the sizing of equipment grounding conductors, as well as equipment bonding jumpers, and, even main bonding jumpers in service equipment, and system bonding jumpers for separately-derived systems, is to calculate the fault current that you expect to flow through these conductors, based on a fault current analysis, using the Eaton SPD Handbook. This method will assure a properly designed grounding and bonding system.

Where ground-fault currents are not excessive, using the conductor sizes expressed in Table 250.122, may be acceptable.

Remember, the object is to establish an 'effective ground-fault current path' (250.4(A)(5)), (250.4(B)(4)).

For main bonding jumpers, system bonding jumpers, and supply-side bonding jumpers for AC systems, refer to Table 250.102(C)(1) for the minimum sizes of these conductors.

Table 250.66 identifies the size of grounding electrode conductors for AC systems. These sizes range from 8 AWG to 3/0 AWG copper or 6 AWG to 250 kcmil aluminum. Once again, these are minimum sizes. In addition to providing protection against lightning and external power faults, this conductor is intended to hold the grounded (neutral) conductor and equipment grounding system at, or near, earth potential ('0' volts). So, the length of this conductor is an important consideration. Once again, basing the data from Table 250.66 on a maximum length of 100 feet (30.48 meters), if the grounding electrode conductor exceeds this length, the conductor size would have to be increased in order to reduce the effects of voltage-drop and to limit the resultant voltage-rise on conductors and equipment. But, as we have stressed in previous sections of this book, the length of the grounding electrode conductor should be limited, due to the moderate to high frequency effects of lightning currents, and the increased impedance of this conductor during these conditions (250.4(A)(1), Informational Note No. 1).

An overload condition may cause dangerous overheating to develop in conductors and equipment. Therefore, the system design must take that into consideration, not only fault conditions, but also overload conditions.

Photovoltaic (PV) System - This system, including all components, is designed to convert solar energy into electric energy. Article 690 applies to Solar Photovoltaic Systems. This Article is quite extensive, so we will identify some important topics.

The System Grounding Requirements for PV Systems Involve the following Grounding Arrangements (690.41)

1. Two-wire PV <u>Arrays</u> with one functional grounded conductor.

   An <u>Array</u> is defined in 690.2 as a mechanically integrated assembly of module(s) or panel(s) with a support structure and foundation, tracker, and other components, as required, to form a DC or AC power producing unit. A <u>module</u> is a complete environmentally protected unit consisting of solar cells, optics, and other compounds, designed to generate DC power when exposed to sunlight. Photons are 'particles of light' that cause electrons to move in a semiconductor. A <u>panel</u> consists of a collection of modules mechanically fastened together, wired, and designed to provide a field-installable unit.

   For example, consider an arrangement of PV modules that generate DC power at the appropriate voltage and current. This is the <u>Photovoltaic Power Source Circuit,</u> and it extends from the power source to a common connection point(s) of the DC system. When there are two or more PV Source Circuits, a <u>Direct Current Combiner</u> is used to combine these circuits and provide for <u>one</u> DC circuit output. The <u>Photovoltaic Output Circuit</u> extends from the PV Source Circuit to the <u>Inverter</u> or to the DC utilization equipment. The <u>Inverter Input Circuit</u> includes the conductors that are connected to the DC input of the inverter. The <u>inverter</u> is the equipment that converts the DC input to an AC output. The <u>Inverter Output Circuit</u> includes the conductors that are connected to the AC output of an Inverter and to the AC connected load.

   A Functional Grounded PV system is one where there is no solidly grounded conductor on the DC side of the Inverter. And the <u>Equipment Grounding Conductor</u> from the Inverter AC output circuit(s) provides the ground connection for the Ground-Fault Protection and equipment grounding of the PV arrays. The <u>EGC</u> is connected to the grounding electrode (system) on the AC side of the Inverter.

2. <u>Bipolar PV arrays</u>, with a functional ground reference (center tap). A Bipolar PV array has 2 DC outputs, each having an opposite polarity to a common reference point or center tap, which is where the ground (earth) connection is made.

3. PV arrays not isolated from the grounded inverter output circuit.

4. Ungrounded PV arrays

5. Solidly grounded PV arrays as permitted in 690.41(B), Exception. In this case, this Exception permits the ground-fault protection to be omitted for PV arrays with not more than 2 PV Source Circuits, and with all PV system DC circuits not on or in buildings.

PV systems that use other methods that accomplish equivalent system protection in accordance with 250.4(A), with equipment <u>listed</u> and <u>identified</u> for the use. Section 250.4(A) covers electrical systems that are grounded (solidly), as well as the bonding of electrical equipment, and a means of providing an effective ground-fault current path, which for a solidly grounded system, will facilitate the operation of the circuit overcurrent device during a ground-fault, or initiate an alarm on a high-impedance grounded system.

Section 690.41(B) requires Ground-Fault Protection for DC PV arrays, with the exception of the PV arrays with not more the 2 PV Source Circuits, where all PV system DC circuits are not on or in buildings and the system is <u>solidly grounded</u>. The Ground-Fault protection must detect ground-faults in the PV array DC conductors and equipment, and the faulted circuits must be automatically disconnected, or the inverter must automatically stop the current flow to the output circuits, as well as isolate the PV system DC circuits from the ground reference in a <u>Functional Grounded System</u>.

The Ground-Fault Protection System is <u>not</u> provided as a means of personnel protection. It is meant to provide protection against fire hazards.

The connection to ground for any current-carrying conductor is made by the Ground-Fault Protective Device for <u>functional grounded systems</u>, or for <u>solidly-grounded PV systems</u>, the DC circuit grounding connection is made at any single point of the PV Output Circuit, that is, from the PV Source Circuit to the Inverter.

Several years ago, PV systems were solidly grounded, that is the DC Source Circuit negative conductor was connected to a grounding electrode (system).

The recommendation was to make this connection to ground at a point that was close to the DC sources (modules) to afford better protection from lightning. This is true for any supply system, DC or AC. Also, keep in mind that the PV source is not a constant voltage source due to variations in the ambient temperature. As the temperature decreases, the supply voltage increases (Table 690.7(A)).

The grounding electrode (system) tends to stabilize the voltage. This is <u>not</u> to say that the earth connection would limit this voltage-rise as the ambient temperature decreases.

The grounding electrode (system) would also be used as a means of grounding the exposed noncurrent carrying metal parts of PV module (support) frames, as well as the metal frames of electrical equipment and conductor enclosures.

The AC supply system from the inverter would also be connected to a grounding electrode (system). This would include the grounded conductor (neutral) and the equipment grounding system.

Of course, both systems, DC and AC, could be connected to the same grounding electrode (system), which would include the grounding electrodes referenced in 250.52(A) and 250.166. If the DC and AC systems are connected to separate grounding electrodes, they must be bonded to limit potential differences between the DC and AC systems (250.50), (250.58).

Also, if the grounding electrode is a single ground rod, pipe, or metal plate, with a resistance-to-ground of over 25 ohms, this grounding electrode must be supplemented by an additional electrode, which may be an additional rod, pipe, or plate (250.53(A)(2)).

However, in more recent times, a <u>Functional Grounded System</u> is the preferred method of system grounding. In this case, the DC system is not solidly grounded and the Equipment Grounding Conductor for the Inverter AC output, which is connected to the grounding electrode (system), either at the Inverter or further downstream, provides the necessary ground reference for the DC Ground-Fault Protection and the equipment grounding of the PV supply system. This includes the metal support structures of the PV arrays and any metallic enclosures that are a part of the DC supply. The Ground-Fault Protection will be part of the Inverter.

In areas of significant thunderstorm activity, it may be beneficial to connect the PV array(s) support structures to an auxiliary grounding electrode (system) as a means of lightning protection (250.52),(250.54). If used, it would not be necessary to bond the auxiliary grounding electrode, as the bonding is already afforded through the equipment grounding system. Or, if the building or structure supports a PV array(s), 690.47(A) requires that the arrays be connected to a grounding electrode system that is installed in accordance with Part III of Article 250 (250.52(A)). For example, if the building or structure has a grounding electrode system consisting of metal in-ground support structures (250.52(A)(2)), the PV array support system will be bonded to this grounding electrode. In this way, the array equipment grounding conductors will have a connection to the grounding electrode system.

Equipment Grounding

Just as in other electrical systems, the exposed noncurrent carrying metal parts of the PV equipment, including the PV module supporting frames, metal enclosures, and metallic raceways and cable assemblies are required to be grounded. Equipment grounding conductors must be installed within the same raceway or cable assembly or otherwise run with the circuit conductors (690.43(C)).

In this respect, this requirement is similar to 300.3(B), even though these circuit conductors from the PV arrays to the Inverter are <u>Direct Current,</u> and not subject to the inductive heating of Alternating Current circuits. And the increased circuit impedance of AC circuits.

The size of the equipment grounding conductors for PV Source and PV Output Circuits are based on 250.122. This means that the EGC is sized in accordance with the size of the circuit overcurrent device.

However, the circuits may originate from the PV modules, where there are no overcurrent devices protecting these circuits. In this case, an assumed overcurrent device, in accordance with 690.9(B), will determine the size of the equipment conductor(s).

Section 690.9(B) requires that the overcurrent devices that are used in <u>PV systems must be listed.</u> This means that these devices must be listed and identified for DC systems, which excludes overcurrent devices that are listed only for AC systems. Alternating current passes through zero two times during each cycle. Direct current is constant, and, especially during short-circuit or ground-fault conditions, these faults are more difficult to interrupt than in alternating current circuits.

For PV systems, the overcurrent devices must be rated at not less than 125% of the maximum currents calculated in accordance with 690.8(A).

Or, if an <u>assembly</u> contains overcurrent devices that are listed for continuous operation at 100% of their rating, these overcurrent devices may be loaded to 100% of their rating (690.6(B)(1), Exception).

For adjustable electronic overcurrent protective devices, where the adjusting means is external to the overcurrent device, the rating of the protection is considered to be the maximum setting possible.

If the adjusting means of the overcurrent device is protected by removable and sealable covers, bolted equipment enclosure doors, or locked doors that are accessible to qualified persons only, the rating of the overcurrent device is considered to be the setting of the adjustment (long-time pickup setting). (690.8(B)(3), (240.6(B),(C))).

690.8(A)(1),(1) states that the maximum calculated current is the sum of the parallel connected PV module rated short-circuit currents multiplied by 125%.

For Example:

10 series connected DC modules with a short-circuit current rating of 8.9 amperes, each.

8.9 amperes × 1.25 (125%) = 11.125 amperes.

The maximum PV Source Circuit Current is 11.125 amperes (690.8(A)(1)).

The overcurrent device rating will be 1.25 (125 %) times 11.125 amperes, or 13.90 amperes to satisfy 690.9(B)(1).

When the 125% value of 690.8(A)(1) is applied to calculate the maximum PV Source Circuit current, and this is combined with the overcurrent device rating of 125% from 690.9(B)(1), the result is 156% (1.25 × 1.25 = 1.5625). 8.9 amperes × 1.5625 = 13.91 amperes

Section 690.9(A) Exception states that overcurrent protection is not required for PV Source Circuits, if the short-circuit currents from all sources does not exceed the ampacity of the DC circuit conductors.

So, if the calculated short-circuit current from 690.8(A) and 690.9(B) is 13.91 amperes, the equipment grounding conductor may be 14 copper (2.08mm²) from Table 250.122. This applies whether, or not, the source circuits are provided with overcurrent protection (690.45).

The equipment grounding conductors for the PV Source Circuits and PV Output Circuits (the circuit(s) from the PV combiner to the Inverter) do not have to be increased in size to compensate for voltage-drop (690.45). These circuits are DC, and are not subject to inductive reactance, as would be the case for AC circuits. The minimum size is No.14 (2.08mm²).

For PV systems with a generating capacity of 100 kW or more, the PV Source Circuits may be determined through calculations made by a licensed professional electrical engineer. An industry standard that may be used to determine the maximum current of a PV system is 'Sand 2004-3535', from Sandia National Laboratories. The use of this calculation method will typically result in a lower current value than the calculation method of 690.8(A)(1),(1). However, the calculation from this standard must be not less than 70% of the value from 690.8(A)(1),(1).

In addition, for grounded DC systems, the Direct-Current System Bonding Jumper, which is used to connect the grounded conductor to the equipment grounding conductor(s), is sized in accordance with the size of the required system grounding electrode conductor (250.166). This DC System Bonding Jumper serves the same purpose as the 'Main Bonding Jumper' for AC systems. It may be installed at the source or at the first system disconnecting means supplied from the source. In this regard, it is the same as a typical Separately-Derived AC System from Section 250.30(A)(1),(250.164(B)).

Any exposed noncurrent carrying metal parts of PV module frames or panels and electric equipment enclosures must be grounded, regardless of voltage (690.43).

Metallic mounting structures (not building steel) may be <u>identified as equipment grounding conductors</u>, and <u>identified</u> bonding jumpers or devices may be used to provide an acceptable bonding means between separate metallic sections. Equipment mounted on these structures (modules) may be bonded with <u>listed devices identified</u> for this purpose. The structure must be bonded to the equipment grounding system (690.43(A),(B)).

Where equipment grounding conductors are installed for the PV arrays and the metal structure used for their support, they must be contained within the same raceway, cable, or otherwise run with the PV Array circuit conductors (690.43(C)).

The grounding of exposed PV module metal frames, conductor enclosures, and metallic equipment enclosures are grounded in accordance with 250.134 and 250.136(A).This equipment is grounded by an equipment grounding conductor, as referenced in 250.118.The equipment grounding conductor is normally run with the circuit conductors for important impedance reduction (300.3(B)),(250.134(B)). For DC circuits, the equipment grounding conductor may be run separately from the circuit conductors, as inductive reactance is not a factor in these circuits. However, 690.43(C) requires the equipment grounding conductors for the PV array and support structures to be contained within the same raceway, cable, or otherwise run with the PV array circuit conductors, when these circuits are extended beyond the vicinity of the array.

For PV module frames, electrical equipment and conductor enclosures, the provisions of 250.136(A) and 250.134 may apply. In this case, the equipment grounding conductor may not be routed with the circuit conductors. And where the equipment grounding conductors are smaller than No. 6 AWG (13.30 mm^2), physical protection, in the form of a raceway or cable armor is required, unless the conductor is installed in such a way that physical protection is not necessary (690.43).

However, 690.43(C) requires the equipment grounding conductors for the PV array to be within the same raceway or cable, or otherwise run with the PV array circuit conductors.

This information is another example of the Code Arrangement of Section 90.3, where Chapters 1-4 apply, except as amended by Chapters 5-6 and 7. Section 300.3(B) requires that conductors of the same circuit, including equipment grounding conductors, must be contained in the same raceway, cable, auxiliary gutter, cable tray, cable bus assembly, cable, or cord in order to limit overall circuit impedance. Section 300.5(I) identifies the same requirement for underground systems.

Grounding Electrode System

The grounding of AC modules, or the grounding of the AC system from an Inverter to a grounding electrode (system) will be in accordance with 250.50 through 250.60. Any ferrous metallic enclosure for a grounding electrode conductor must be bonded (on both ends) to the internal grounding electrode conductor to comply with 250.64(E)(1).

For solidly grounded systems, it is common to have photovoltaic systems with both DC and AC grounding requirements. The DC grounding system must be bonded to the AC grounding system in order to limit voltage differences between these systems. This is a similar requirement to 250.50 and 250.58, and for the same reason. If two grounding electrode conductors are installed, one for the DC system and one for the AC system, a bonding conductor may be installed between these systems. The bonding conductor size is based on the larger size of the AC grounding electrode conductor or the DC grounding electrode conductor (250.66),(250.166).

Or, a DC grounding electrode conductor, sized in accordance with 250.166, may be installed from the DC grounding electrode connection point to the AC grounding electrode.

It is also common on PV systems to have an isolation transformer installed in order to separate the DC grounded circuit conductor from the AC grounded circuit conductor. Once again, there must be a bonding connection between these systems. Or, these systems may be connected to a common grounding electrode in order to establish the same zero-volts potential reference to the earth.

Any auxiliary grounding electrode (Section 250.54) installed at ground level for roof-mounted or pole-mounted arrays must be connected to the array frame or supporting structure. It may be possible to bond the frame of a roof-mounted array to the metal frame of a building, if the metal building frame is part of the grounding electrode system (250.52(A)(2)), (690.47(B)).

It should be noted that the continuity of the equipment grounding system must be maintained, even if equipment is removed for repair or replacement. The PV source and PV output circuits remain energized as long as the PV modules are exposed to light. For example, if the inverter is removed for service, a bonding conductor must be installed to maintain the connection to the grounding system.

In addition, if the removal of equipment causes the Main Bonding Jumper in the Inverter to be disconnected, a bonding jumper must be installed to assure the grounding connection to the system grounded conductor. There may be a significant voltage-rise on the grounded conductor if the connection to the grounding electrode is interrupted.

A 'Functional Grounded PV System' has an electrical reference to ground through an element (fuse, circuit breaker, or electronic device) that is a part of a listed ground-fault protection system, and that is not solidly grounded. These systems will be at earth potential under normal conditions, but may be at an elevated voltage above earth potential during fault conditions.

*Plenum-

A plenum (chamber) (NFPA 90A) typically, is the collection point for the air distribution system. One or more air ducts are connected to the plenum for the purpose of transporting environmental air.

Section 300.22(B) covers the wiring methods permitted in ducts specifically fabricated for 'environmental air.' These wiring methods are limited, due to the concern of producing toxic smoke in a fire environment. Only metallic type wiring methods are permitted. These include MI cable, MC cable, electrical metallic tubing, flexible metallic tubing, intermediate metal conduit, and rigid metal conduit. These cables and raceways must not have a nonmetallic covering. Where flexibility is required, flexible metal conduit in lengths up to 4 feet (1.2 meters) is permitted.

Enclosed and gasketed luminaires are permitted to facilitate maintenance.

Electrical equipment and these associated wiring methods are permitted only if they are used for the purpose of sensing the environmental air, and, or, to cause the operation of specialized devices, such as fire dampers.

Section 300.22(C) (NFPA 90A)

This section covers the air-handling space above a suspended ceiling used to transport the return air of an air distribution system. The ceiling cavity is, in effect, a 'plenum' by definition, even though this space was not fabricated for this use.

Once again, the wiring methods that are permitted in this air-handling space are limited to metallic types, or those nonmetallic types that have recognized 'low-smoke' and 'low heat' release properties.

Permitted wiring methods include electrical metallic tubing, flexible metallic tubing, intermediate metal conduit, rigid metal conduit with no nonmetallic covering, flexible metal conduit, surface metal raceway, metal wireway with metal covers, <u>MI cable. MC cable, AC cable (without a nonmetallic covering</u>), and control or power cables specifically listed for air-handling spaces.

In addition, cable ties and other cable accessories must have 'low-heat' and 'low-smoke' producing characteristics in order to be used in air-handling spaces (UL 2043).

Chapters 7 and 8 include references to 'Plenum Rated' Cables that are specifically permitted in 'Ducts Used for Environmental Air' (300.22(B)), and 'Other Space for Environmental Air' (300.22(C)).

As we begin our discussion about the types of cables that are suitable for 'General Purpose Use,' cables that may extend from floor to floor in a building (Riser Cables), and those that are suitable in air-handling spaces (Plenum Rated Cables), we should identify the testing criteria for these cables.

The first test is the 'Vertical Tray Flame Test' (UL-1685). This test is performed by installing the cable in a 5 foot (1.53 meters) vertical section of cable tray, with the cable ignited at the bottom and flame travel monitored for 20 minutes. The 'char length' is limited to 4 feet, 11 inches (1.5 meters) for this 20 minute period. This test does not monitor the smoke-producing characteristics of the cable. Cables that have passed this test would be suitable for general purpose applications, and would not be acceptable in air-handling spaces.

In Chapters 7 and 8 the following cable types would be suitable for this purpose:

Class 2 or Class 3 Cables

Section 725.179(C) - CL 2 and CL 3

Fire Alarm Cables

Section 760.53-NPLF
Section 760.135-FPL
Table 760.154

Table 760.154(A)
Table 760.176(G)

Optical Fiber Cables

Section 770. - OFN-OFC
Table 770.154(a)
Table 770.154(b)
Table 770.179

Communications Cables

Section 800.113 CM
Table 800.154(a)
Table 800.154(b)
Table 800.154(c)
Table 800.179

CATV Cables

Section 820.113 CATV
Table 820.154(a)
Table 820.154(b)
Section 820.179
Table 820.179

Cables that are suitable for extending from floor to floor in a building will have a 'Riser' rating (UL 1666). This test, once again, monitors flame propagation, but not the smoke producing characteristics of the cable. The cable has a low-flammability, in that the cable will burn, but will self-extinguish when the flame is removed. Toxic and corrosive gases are released during the test. So, the cable is not suitable for use in air-handling spaces. Suitability is limited to use inside walls and vertical shafts in buildings. Cables that have passed this test are identified with the letter 'R' as a suffix (CL2R-CL3R), (NPLFR-FPLR), (OFNR-OFCR), (CMR), (CATVR).

Cables that may be installed in limited lengths in 'Ducts Specifically Fabricated for Environmental Air' and 'Other Spaces Used for Environmental Air (Plenums)' (300.22(B) -300.22(C)), would have a 'Plenum' rating where they are exposed in these spaces.

Plenum cables have a low-flammability, in that the cable will burn, but will self extinguish when the flame is removed. Toxic and corrosive gases will be released when the cable is burned, but the smoke density is limited. (UL 910-NFPA 262).

Cables with a 'Plenum' rating will be identified with the letter 'P' as a suffix. (CL2P-CL3P), (NPLFP-FPLP), (OFNP-OFCP), (CMP), (CATVP).

However, the wiring methods that are suitable for use in ' other spaces used for environmental air ', including Plenum Rated cables, are only permitted to connect to equipment or devices that are associated with the direct action upon or sensing of the contained air.

The total length of this wiring cannot exceed 4 feet (1.2m),(300.22(B)), Exception.

*Premises Wiring (System)-

Certainly this wiring system has expanded with the inclusion of additional power sources, and not only from the utility source. As the definition implies, the premises wiring includes power, lighting, control, and signal wiring, along with the associated hardware, fittings, and wiring devices that are temporarily or permanently installed.

For service supplied systems, the 'Service Point' establishes the beginning of the premises wiring. And, it extends to the farthest outlet on the distribution system. The 'Service Point' is the physical point of connection between the utility and the customer premises wiring system.

Or, the premises wiring begins at another power source (solar photovoltaic systems, batteries, generators, etc.), and extends to the outlets.

The premises wiring does not include the internal wiring of equipment.

*Pressurized as Applied to Hazardous (Classified) Locations-

The process of supplying an enclosure with a protective gas, with or without continuous flow, and at sufficient pressure, to prevent the entrance of combustible dust or ignitable fibers or flyings.

*Process Seal as Applied to Hazardous (Classified) Locations-

A seal between electrical systems and flammable or combustible process fluids where a failure could allow the migration of process fluids into the premises wiring system (501.17).

*Purged and Pressurized as Applied to Hazardous (Classified) Locations-

The process of (1) purging, that is, supplying an enclosure with a protective gas at a sufficient flow and positive pressure to reduce the concentration of any flammable gas or vapor initially present to an acceptable level; and (2) pressurization, supplying an enclosure with a protective gas with or without continuous flow, at sufficient pressure to prevent the entrance of a flammable gas or vapor, a combustible dust, or an ignitible fiber (ANSI/NFPA 496).

NFPA 496 requires that the enclosure is purged with a minimum of 4 volumes of air/inert gas before any circuits or equipment are energized. A positive pressure of a minimum of .1 inch of water must be maintained, with or without continuous flow.

Type X reduces the Classification from Division 1 and Zone 1, to Unclassified.

Type Y reduces the classification from Division 1 to Division 2, or Zone 1 to Zone 2

Type Z reduces the Classification from Division 2 or Zone 2, to Unclassified.

*Qualified Person-

One who has demonstrated skills and knowledge related to the construction and operation of electrical equipment and installations and has received safety training to identify the hazards and reduce the associated risk (NFPA 70E-Article 100).

NFPA 70E Article 120 applies to the process of creating an electrically safe work condition and Article 130 defines the situations under which an electrically safe work condition must be established.

*Raceway-

An enclosed channel designed for the containment of wires, cables, or busbars with additional functions, such as, physical protection, and, usually, if of metal construction, as an equipment grounding conductor (250.118).

Chapter 3 is replete with examples of various types of metallic and nonmetallic raceways. Section 250.118 identifies the types of metal raceways that are suitable for use as equipment grounding conductors, sometimes with certain restrictions. If the metal raceway is to be used for this purpose, other conditions should also be considered, such as in an installation in a wet or corrosive environment, where threaded joints may be affected over time. The length of the raceway is another concern. Just as an equipment grounding conductor, in the form of a wire, has restrictions to its length when serving as an effective ground-fault current path due to voltage-drop (and the resultant voltage-rise on the frames of equipment), the length of the metal raceway should be a consideration for the same reason.

Threaded metal conduits must be made wrenchtight to assure electrical continuity and to comply with 300.10. However, it is normal for relaxation to occur over time.

Also, for installations of flexible metal conduit in a Class I, Hazardous (Classified) Location, an equipment grounding (bonding) conductor must supplement the flexible metal conduit to assure electrical continuity and to prevent arcing or sparking at terminations due to poor connections (501.30(B)). The size of the bonding jumper is in accordance with 250.102(D) and 250.122, on the load side of an overcurrent device..

This restriction also applies to Liquidtight Flexible Metal Conduit in Class II and Class III Hazardous (Classified) Locations in Sections 502.30(B) and 503.30(B).

In addition, Section 517.13(B) requires a redundant equipment grounding conductor (insulated, not bare) as a supplement to a metal raceway for the purpose of grounding the metal boxes and enclosures that contain receptacles, for grounding the metal enclosures of fixed electrical equipment, and for connection to the grounding terminals of all receptacles in a patient care area of a Health Care Facility.

One final note, Cable Tray (Article 392) is a wiring support system, and, even if provided with covers, it is not a raceway. However, metallic cable trays may be used as equipment grounding conductors. The joints and discontinuous segments must be properly bonded to assure the equipment grounding conductor integrity (392.60(A)(B)), (250.96). If the cable tray is provided with rigidly bolted joints, a bonding jumper is not required. Where bonding is required, such as at discontinuous joints, the bonding jumper must be sized and installed in accordance with 250.102(D) and 250.122.

There are many raceway types referenced in Chapter 3. The most common types are rigid metal conduit, intermediate metal conduit, electrical metallic tubing, and rigid polyvinyl chloride conduit (PVC). There are common requirements for these raceways, such as reaming to remove rough edges after cutting, and where the IMC or RMC are threaded in the field, a standard cutting die (¾ inch taper per foot) must be used. In each of these raceways, the number of conductors are limited by the fill percentages in Table 1, Chapter 9. This also applies to the number of cable assemblies within these raceways.

Table 1

Number of conductors and/or cables	Cross-sectional area %
1	53
2	31
Over 2	40

Where conduit or tubing nipples have a length not exceeding 24″ (600mm), the fill percentage may be 60% of the cross-sectional area of the raceway.

The ampacity adjustment factor from 310.15(B)(3)(a) (proximity effects) does not apply to these conduit nipples.

Equipment grounding or equipment bonding conductors are included in these calculations for percent fill.

For cable assemblies, including optical fiber cables, the cable diameter is used to calculate percent fill.

If the cable has an elliptical cross-section, the major diameter of the ellipse is considered to be the cable diameter, and the cross-sectional area is calculated accordingly. For example, an elliptical cable with a widest point measurement of 2.5″ (63.5mm),(Chapter 9-Note 9).

$$area\ of\ circle\ -\pi r^2 = 3.1416 \times 1.25''^2$$

$$
\begin{array}{cc}
1.25 & 3.1416(\pi) \\
\times 1.25 & \times 1.5625 \\
\hline
1.5625 & 4.90875\,sq.in.
\end{array}
$$

If this was a single cable assembly, installed in a rigid metal conduit, Table 4 of Chapter 9 would require a 3½″ trade size conduit, (5.305 sq. in-1 conductor –53% fill).

If single conductors of different sizes and/or insulation types are to be installed in these conduits, the calculation is to be made in accordance with Table 5 of Chapter 9, and the conduit is sized from Table 4.

*Rainproof- Raintight-

This type of construction coincides with Table 110.28.

Informational Note No. 1 identifies 'Raintight' enclosures as NEMA 3, 3S, 3SX, 3X, 4, 4X, 6 and 6P.

Rainproof enclosures would include NEMA 3R and 3RX.

*Receptacle-

This may be a single or multiple contact device (duplex, triplex, etc.) installed at an outlet and serves to accommodate an attachment plug(s). Article 406 addresses Receptacles, Cord Connectors, and Attachment Plugs.

*Separately-Derived System-

This term has the potential for many applications in an electrical distribution system. It is very often a transformer, but, it may be a generator, batteries, solar photovoltaic system, or wind electric system.

This system may be supplied from the service equipment. For example, a feeder extended from the service equipment that supplies the primary of a transformer. The secondary of this transformer establishes the beginning of the 'separately-derived system.'

Or, an onsite generator that supplies a building or structure, or possibly, a specific piece of equipment.

Or, a solar photovoltaic system that supplies a limited AC load.

Section 250.30 addresses the requirements for grounding separately-derived systems. This includes single separately-derived systems and multiple separately-derived systems (250.30(A)(5), (250.30(A)(6).

So, let's begin here. If the separately-derived system is a grounded system, the grounded conductor will be bonded to the equipment grounding system through a bonding conductor, which is a System Bonding Jumper. This bonding conductor serves the same purpose as a Main Bonding Jumper in the service equipment, and it is sized in the same way (250.102(C)(1)). The bonding connection may be established at the source of the separately-derived system, or, at the first disconnecting means or overcurrent device that is supplied from the source. As a matter of fact, if the separately-derived system is an outdoor transformer, there may be a system bonding jumper at the source, as well as at the first disconnecting means. However, if this is done, there can be no conducting path that parallels the grounded conductor, which may be a metal raceway that contains these supply conductors, or, an equipment grounding conductor which is run with the supply conductors. If this were done, an 'objectionable current' would flow over these conducting paths. For example, the normal current returning to the system source through the grounded conductor would also flow through the downstream 'System Bonding Jumper' at the first disconnecting means or overcurrent device, and then return to the source through the metal raceway or through the equipment grounding conductor. Of course, this current should not be flowing through this equipment grounding conductor. And it is this current flow that is considered 'objectionable', according to Section 250.6(A).

The reason that an option is given to connect the grounded conductor and the equipment grounding conductor at the source of the separately-derived system, or at the first disconnect or overcurrent device supplied from the source, is that it is from this connection that the grounding electrode conductor would extend and connect to the grounding electrode. In order to limit the length of the grounding electrode conductor, so that the grounded conductor and the equipment grounding conductor are held at, virtually, earth potential ('0' volts), the location of the grounding electrode would determine whether this connection should be at the source, or at the first disconnect or overcurrent device. In other words, if the grounding electrode was closer to the source of the separately-derived system than it was to the first disconnect or overcurrent device, the bonding and grounding electrode conductor connection should be made there. If the grounding electrode were closer to the first disconnecting means or overcurrent device than it was from the source of the separately-derived system, then the bonding connection and the grounding electrode conductor connection should be made there (250.30(A)(1)), (See 250.164(B) for a DC-on Premises System).

The grounding electrode will be one, or more, of the grounding electrodes referenced in 250.52(A).

The grounding electrode conductor is sized from Section 250.66, based on the largest size of the ungrounded conductors, and a reference to Table 250.66. However, for connections to a ground rod, pipe, or plate, a No. 6 AWG (13.30 mm²) copper conductor may be used.

For a connection to a 'ground ring', a No. 2 AWG (33.63mm) copper conductor may be used.

For a concrete-encased electrode, a No. 4 AWG (21.15 mm²) copper conductor may be used.

Another important issue must be considered, and that is the length of the ungrounded conductors that extend from the source of the separately-derived system to the location of the first overcurrent device

For example, if the separately-derived system were supplied from a transformer secondary, the unprotected secondary conductors would comply with Section 240.21(C).

In the first case, (240.21(C)(1)), the transformer secondary conductors of a single-phase system, having a 2-wire, single-voltage secondary (e.g., 240/120 volts), may be protected by the transformer primary overcurrent device, providing that the primary overcurrent device has been properly sized (450.3(B)), and this primary overcurrent device has a rating that does not exceed the value that is determined by multiplying the ampacity of the secondary conductors by the <u>secondary</u> to <u>primary</u> transformer voltage ratio.

For example, a single-phase, 240/120 volt, 10 kVA transformer with a full-load primary current of 41.67 amperes. From Section 450.3(B), the full-load primary current may be increased by 125% (1.25), or 52 amperes. (41.67 × 1.25 = 52 amperes). If we decide to use the next larger size overcurrent device, it would be 60 amperes (450.3(B), Exception No. 1).

The secondary-to-primary transformer voltage ratio is 1/2 (<u>120/240V.)</u> or .5

$$\frac{60 \text{ amperes}}{.5} \text{ (primary overcurrent device(s)} = 120 \text{ amperes}$$
$$\text{(secondary-to-primary voltage ratio)}$$

Therefore, if the secondary conductor ampacity was at least 120 amperes, these conductors would be considered to be protected by the primary 60 ampere overcurrent device(s).

These provisions would also apply to a 3-phase, 3-wire, Delta-to-Delta connected transformer.

In these two cases, no overcurrent protection is required for the transformer secondary conductors because short-circuits and ground-faults are effectively transferred to the transformer primary to cause the primary overcurrent devices to operate.

This is not the case for other transformer connections.

For example, in a 3-phase, 4-wire, Wye connected transformer secondary supplied from a 3-phase Delta connected primary, there is a 30-degree phase-shift from primary to secondary, that is the primary voltage will lead the secondary voltage by 30 degrees.

This produces a condition where faults on the secondary may not be effectively transferred to the primary in order to operate the primary overcurrent device(s).

Also, where the secondary is a single-phase, 3-wire system (e.g. (480/240/120 volts), the transformer secondary conductors require their own properly sized secondary overcurrent device, irrespective of any overcurrent protection on the primary side.

This also would apply to a 3-Phase, Delta-to-Delta system, with a 4-wire secondary.

And the location of the secondary overcurrent device will be in accordance with the following Sections.

Section 240.21(C)(1) - As we have already discussed, the transformer secondary conductors of a single-phase, two-wire system may be considered as being protected by the transformer primary overcurrent device. And this protection is based on the ratio of the secondary to primary voltage.

$$(e.g., \frac{120 \text{ v}}{480 \text{ v}} = 1/4, \text{ or, } .25 - \frac{240 \text{ v}}{480 \text{ v}} - 1/2, \text{ or } .5)$$

The transformer secondary conductor ampacity is multiplied by this ratio and the ampere rating of the primary overcurrent protection must not exceed this value.

This protection scheme would also apply to 3-phase, 3-wire systems. And, this provision is referenced in Section 240.4(F).

Section 240.21(C)(2) - This is a transformer tap rule, where the secondary conductors are not longer than 10 feet (3 meters) in length.

These conductors are properly sized in accordance with the connected load, and these conductors have an ampacity of not less than the overcurrent device in the equipment that they are supplying, or, not less than the rating of the overcurrent device at their termination.

A Note here references an exception where these tap conductors supply listed equipment that specifies the minimum conductor size.

The tap conductors do not extend beyond the equipment that they are supplying.

These conductors are provided with physical protection in the form of a raceway.

Finally, where the tap conductors extend beyond the equipment where the tap is made, the overcurrent device protecting the transformer primary has a rating which does not exceed 10 times the ampacity of the secondary conductors, when multiplied by the secondary-to-primary transformer voltage ratio.

Example

A 50 kVA Transformer- 3-Phase
480 volt primary
208/120 volt secondary

Full-load Primary Current - $\dfrac{50,000va}{831.36 \ (480 \ v. \times 1.732)} = 60.142$, or

60 amperes
(Section 220.5(B) (Informational Annex D)
(Fractions of an ampere less than .5 may
be dropped)

Rating of Primary Overcurrent Device -

60 amperes
$\times 1.25$ (Table 450.3(B)
75 amperes (Table 450.3(B) - Note 1 gives permission to round-up to the next standard size, or 80 amperes).

Secondary-to-Primary Transformer Voltage Ratio

$$\frac{208 \ \text{Volts}}{480 \ \text{Volts}} = 0.433$$

The transformer secondary conductors must have an ampacity, that when multiplied by the secondary-to-primary transformer voltage ratio, will be at least 10% (1/10th) of the rating of the primary overcurrent device, or 80 amperes, in this case.

$$\frac{80 \text{ amperes}}{0.433} = 184.757$$

184.757 amperes
$\times .10$ (10 percent)
18.4757 amperes, or 18 amperes

Or,

This method may be used for this calculation.
Primary to Secondary Transformer Voltage Ratio $\frac{480 \text{ Volts}}{208 \text{ Volts}} = 2.31$
Rating of Primary Overcurrent Device - 80 amperes = 8 amperes
(1/10th of Primary Overcurrent Device Rating)
(80 a $\times .10 = 8$ amperes)
8 amperes
$\times 2.31$
18.48 amperes, or 18 amperes (220.5(B))

The transformers secondary conductors must have an ampacity of at least 18 amperes.

A 12 AWG copper conductor would be acceptable for these secondary conductors, which has an ampacity of 20 amperes from the 60°C. Column of Table 310.15(B)(16). This wire size would correlate with the temperature limitations of Section 110.14(C)(1),(a).

Once again, I would like to point out that this limited length (10 feet - 3 meters) tap would not be permissible in other applications (240.4(F)). The protection of transformer secondary conductors by the transformer primary overcurrent device is limited to single-phase, 2-wire transformer secondary systems, or, 3-phase, 3-wire, single-voltage systems. The transformer in our example is a 3-phase, Delta-to-Wye system (480V.-208/120V.). The secondary is a 4-wire, Wye system. There is a 30 degree phase shift in this delta-to-wye connection, as we have mentioned earlier. This means that short-circuits and ground-faults on the secondary side are not effectively transferred to the primary. For each one ampere of fault current on the secondary side of the transformer, the primary sees .58 amperes. So, typically, the secondary conductors would require their own overcurrent protection, irrespective of the primary overcurrent devices.

However, this is a tap rule and the provisions of this tap rule specifically state that the secondary conductors have an overcurrent device at the termination of these conductors. And due to the limited length of the secondary conductors, and, the fact that they are provided with physical protection in the form of a raceway, it is less likely that physical damage to these conductors would lead to short-circuits or ground-faults. But, it is certainly a possibility. And if a fault does occur in the secondary, whether it happens in the secondary winding, or in the secondary tap conductors, or in equipment supplied by these conductors, the rapid clearing of the primary overcurrent device is questionable.

240.21(C)(3) - This transformer tap rule permits the transformers secondary conductors to be up to 25 feet (7.5 meters) long. It is permissible to be used in industrial installations with qualified maintenance personnel.

The ampacity of the secondary conductors must not be less than the secondary current rating of the transformer, and the rating of the secondary overcurrent device(s) may not exceed the secondary conductor ampacity.

Physical protection (raceway, or an approved enclosure) must be provided for the secondary conductors.

Section 240.21(C)(4) - This is an unlimited length tap rule, where the secondary conductors are located outside of a building or structure.

There is no length restriction for these secondary conductors. The secondary conductors terminate in an overcurrent device that does not exceed the ampacity of the secondary conductors. This secondary overcurrent device(s) is an integral part of a disconnecting means (fusible switch or a circuit breaker). The location of this disconnecting means may be outside of the building or structure, or inside, nearest the point of entry of the conductors (225.32), (230.6). And this disconnecting means must be readily accessible.

Note: The Authority Having Jurisdiction will normally decide on the acceptable length of the unprotected feeder conductors inside the building.

Section 240.21(C)(5). This tap rule applies where the transformer primary is tapped to a feeder and it correlates with Section 240.21(B)(3). The ampacity of the primary conductors is at least one-third of the ampere rating of the overcurrent device protecting the feeder conductors.

The secondary conductors have an ampacity that when multiplied by the primary-to-secondary voltage ratio is at least 1/3 (one-third) of the rating of the overcurrent device protecting the primary of the transformer.

Example

A 75kVA Transformer, 3-phase, 480 Volt primary and a 208/120 Volt secondary.
Primary full-load current rating $\frac{75,000VA}{831.36\ (480\ V. \times 1.732)} = 90.21$ amperes

This transformer is tapped to a feeder that is protected at 250 amperes, which is not more than three times the primary full-load current rating.

Primary Conductors

From Table 310.15(B)(16), the primary conductors are No. 2 AWG copper $(33.63mm^2)$ 60°C. = 95 amperes).

Secondary Conductors

The primary-to-secondary transformer voltage ratio is $\dfrac{480\ V}{208\ V} = 2.31$

$$\dfrac{250A.}{2.31} = 108.23 \qquad\qquad \begin{array}{r} 108.23 \\ \times\ 3 \\ \hline 35.72\ \text{Amperes} \end{array}$$

Secondary conductors – 8 AWG - Copper – 40 amperes @ 60°C.

$$\begin{array}{r} 40\ \text{amperes} \\ \times\ 2.31 \\ \hline 92.40 \\ \times\ 3 \\ \hline 277.2\ \text{amperes} \end{array}$$

The secondary conductor ampacity is at least 1/3 of the rating of the primary overcurrent device, when multiplied by the primary- to-secondary transformer voltage ratio.

Section 240.21(C)(6) - This is another transformer tap rule which permits the secondary conductors to be up to 25 feet (7.5 meters) in length.

The secondary conductor ampacity, when multiplied by the primary-to-secondary transformer voltage ratio, must be at least one-third of the rating of the transformer primary overcurrent device.

The secondary conductors must terminate in an overcurrent device that has a rating that does not exceed the secondary conductor ampacity. And, the secondary conductors are provided with physical protection, such as a raceway or approved enclosure.

Example

3-Phase Transformer - 112.5kVA-480/208/120V.
Full-load Primary current-135.32 amperes

$$\begin{array}{r} 135.32\ \text{amperes} \\ \times 1.25 \\ \hline 169.15\ \text{amperes or 175 amperes} \end{array}$$
(Table 450.3(B), Note1),(240.6(A))

$$\frac{480V.}{208\ V.} = 2.31 \quad -175A.\text{-Primary Overcurrent Device Rating}$$

$$\frac{175A.}{2.31} = 76\ \text{Amperes} \qquad \begin{array}{r} 76\ \text{Amperes} \\ \times\ .33 \\ \hline 25\ \text{Amperes} \end{array}$$

Secondary conductors – 10 AWG-Copper-30 amperes @ 60°C.

$$\begin{array}{r} 30\ \text{amperes} \\ \times\ 2.31 \\ \hline 69\ \text{amperes} \\ \times\ 3 \\ \hline 207\ \text{Amperes} \end{array}$$

The Secondary conductor ampacity is at least 1/3 of the rating of the primary overcurrent device, when multiplied by the primary-to- secondary transformer voltage ratio.

This is a detailed analysis of conditions where conductors are connected to a transformer secondary, and the permissible length of these conductors to the location of the secondary overcurrent device. Section 240.21(C)(1) through (6) has a direct connection to the defined term 'Separately-Derived System', as the transformer secondary establishes the beginning of the Separately-Derived System.

However, when dealing with other systems, such as generators, batteries, solar photovoltaic systems, etc., that may also be supplying Separately-Derived Systems, it is a concern as to the location of the overcurrent device on the load side of these supply systems. The unprotected length of these conductors should be limited.

Generators typically have an overcurrent device integral with the generator. And solar photovoltaic equipment has overcurrent protection, both DC and AC, as part of this equipment. If other types of equipment supply Separately-Derived Systems, the overcurrent protection on the load side of this equipment should be as close as practicable to the source (240.21(G),(H)).

*Service-

These are the conductors and equipment for delivering electric energy from the serving utility to the wiring system of the premises served.

Article 230 covers the requirements that apply to the number of services that may supply a building or structure (normally, only one (230.2), the minimum size of service conductors for overhead installations (8 AWG-8.37mm² copper), (6 AWG-13.30mm² aluminum, or copper-clad aluminum) (230.23(B)), or for underground installations (8 AWG-8.37mm²copper), (6 AWG-13.30mm² aluminum, or copper-clad aluminum) (230.31 (B)).

There is an exception for service conductors that supply limited loads of a single branch circuit, where these conductors may be 12 AWG copper (3.31mm²), or 10 AWG (5.26mm²) aluminum, or copper-clad aluminum (230.23(B), Exception).

At the end of this section, we will analyze a dwelling unit service (standard calculation, optional calculation), a multifamily dwelling, and a commercial (store building) service. Each set of service-drop conductors, each set of overhead service conductors, and each set of underground service conductors, or service lateral, may only supply one set of service-entrance conductors.

Service-Entrance conductors may be derived from an overhead system or from an underground system, unless the service equipment is outside the building or structure, in which case there may not be any service-entrance conductors.

For an overhead system, these conductors connect to the service-drop or overhead service conductors, and they terminate at the service equipment.

For an underground system, these conductors connect to the underground service lateral or underground service conductors, and they terminate at the service equipment (230.40).

The service-entrance conductors must have an ampacity in accordance with the connected load.

Any correction factor for ambient temperature or adjustment factor for the number of current-carrying conductors installed together (proximity effects), must be considered, as well as the effects of continuous loading, when determining the size of the service-entrance conductors.

The grounded conductor may not have to be increased in size to compensate for continuous loading because it does not normally connect to an overcurrent device (230.42(A), Exception No.1).

And, just as in the case for feeder conductors, if the service-entrance conductors terminate in an overcurrent device that is part of an assembly that is listed for operation at 100% of its rating, the effects of continuous loading is not a consideration (230.42(A), Exception No. 2).

Service equipment that is rated 1000 volts or less must be identified as being suitable for use as service equipment and listed or field labeled. And, in accordance with 110.24(A), (B), in other than dwelling units, the service equipment must be legibly marked in the field with the maximum available fault (short-circuit) current.

The service disconnecting means, which may consist of not more than 6 switches or sets of circuit breakers, grouped together, and marked to identify the load served (230.71, 230.72), is to be in a readily accessible location, either outside of the building or structure, or inside nearest the point of entrance of the service conductors.

Normally, local codes, or the authority having jurisdiction, will determine the maximum length that the service conductors may extend into the building before they terminate into the service disconnect (230.70).

Ungrounded service conductors are required to have <u>overload</u> protection and must be provided with an overcurrent device in series with each ungrounded conductor. The rating or setting of the overcurrent device(s) must not be higher than the ampacity of the service conductors (230.90(A)).

These conductors are not provided with ground-fault and short-circuit protection on the line-side of the service disconnect where the transformer is utility owned, unless the customer provides this protection. And, where supplied from a transformer, the only protection is through the transformer primary overcurrent protection. And, because the transformer may be owned by the serving utility, the provisions of 450.3(A), do not apply.

230.82 –There is a limited amount of equipment permitted on the supply side of the service disconnecting means. This makes sense, as this equipment will typically have no means of overcurrent protection, except, possibly, protection provided upstream by the serving utility.

As an addition to the 2020 NEC, single-family and two-family dwellings require a readily accessible external disconnecting means, which will be on the supply side of the service disconnecting means, and this will serve as an aid for fire and rescue personnel (230.85).

The list of this equipment includes:

1. Cable limiters or other current-limiting devices. Cable limiters are effective in isolating a faulted cable, especially where the supply circuit conductors are in parallel.
2. Meters and meter sockets where the voltage does not exceed 1000 volts.
3. Meter disconnect switches rated at up to 1000 volts. These switches must have a short-circuit current rating at least equivalent to, or greater than, the available short-circuit current.
4. Current and voltage transformers, impedance shunts, load management devices, surge-arresters and Type 1 surge-protective devices (285.23(A)(1)).
5. Where provided with a separate disconnect and overcurrent protection, emergency lighting, fire pumps, fire alarm systems, standby power, and sprinkler alarms.

6. Alternate power systems, including solar photovoltaic systems, fuel cell systems, wind electric systems, energy storage systems, or interconnected electric power production sources.
7. Where provided with proper overcurrent protection, control circuits for power operable service disconnecting means.
8. Ground-fault protection systems and Type 2 surge-protective devices that are a part of listed equipment and provided with proper overcurrent protection and a disconnecting means (285.24(A))
9. Connections for the supply of listed communications equipment that is under the exclusive control of the serving electric utility. These circuits must be provided with proper overcurrent protection.
10. A readily accessible disconnecting means for one and two family dwelling occupancies (230.85).

Single Family Dwelling

Floor Area

First floor	1200 square feet
Second floor	1000 square feet
Basement-possible conversion to living space	800 square feet
Total	3000 square feet

$$\frac{3000 \text{ square feet}}{.093} = 279 \text{ square meters}$$

Load

Range	=	12 kw
Dryer	=	5 kw
Water heater	=	2.5kw
Electric-Space heating	=	12kVA–100%–220.51
Air-conditioner	=	4.5kVA-Omit-220.60
Dishwasher	=	960 VA(120 volts)
Disposer	=	720 VA(120 volts)

General Lighting Load

$$\begin{array}{r} 3000 \quad square\,feet \\ \times \quad 3VA \quad (Table\,220.12) \\ \hline 9000\,VA \end{array}$$

3000VA- 100% Remainder @35%
        9000VA   *3000VA*
        3000VA   *3675VA (10,500VA @35%)*
        <u>*1500VA*</u>   *6675VA – Net Total*
       13,500 VA

Net total	−6675 VA
Range	−8000 VA (Table 220.55-Col.C)
Dryer	−5000 VA (Table 220.54)
Water Heater	−2500 VA
Electric Space Heating	−12000 VA(220.51)
Air Conditioning	−0 VA(omit-220.60)
Dishwasher	−960VA
Disposer	<u>− 720 VA</u>
	35,855 VA Total

$$\frac{35,855 \text{ VA}}{240 \text{ Volts}} = 149.40, \text{ or } 149 \text{ amperes } (220.5(B))$$

Neutral Load

6675VA-Lighting , Small Appliance Load, and Laundry Load
5600VA -8000VA @0.70-(220.61 (B)(1))
<u>3500VA</u> -5000VA @ 0.70-(220.61(B)(1))
15,775VA-Total

$$\frac{15,775VA}{240VA} = 65.73, \text{ or } 66A- (220.5(B))$$

Line 1	N	Line 2	
149 A	66 A	149 A	
8 A	8 A		(dishwasher)
	6A	6A	(disposer)
	2A	2A,	(25% of largest motor (430.24))
157A	82A	157A	

Service Rating – 175 Amperes
310.15(B)(7)(1)-Service conductors are permitted to have an ampacity of 83% of the service rating for a one-family dwelling.

        175 Amperes
       <u>× .83</u>
    145.25 Amperes or 145 Amperes

1/0 copper-150 amperes-75°C.-110.14(C)(1)(b)(1)(a)

This Service is rated at 175 amperes.

Neutral   82 amperes

Service conductors-1/0 copper -150 amperes -75°C.-110.14(C)(1) (b),(1),(a)

Service conductors-3/0 aluminum -75°C. -155 amperes-10.14(C)(1)(b),(1),(a)

Neutral conductor -3 AWG-copper-85amperes

Neural conductor -3 AWG-copper-85amperes

Optional Calculation

Single Family Dwelling

Total floor area – 2000 square feet (finished)
Cellar-800 square feet

$$
\begin{array}{r}
2000 \;\; \textit{square feet} \\
\underline{800} \;\; \textit{square feet} \\
2800 \;\; \textit{square feet}
\end{array}
$$

Load

12 – kVA   - Range
10 – kVA   - Electric space heat (5 separate units)
5.5 – kVA   - Dryer
2.5 – kVA   - Water heater
1.5 – kVA   - Dishwasher
10 – amp   - 240 volt air-conditioner

Table 220.12 (General Lighting Load)

$$
\begin{array}{r}
2800 \;\; \textit{square feet} \\
\underline{\times\ 3} \;\; \textit{VA per square foot} \\
8400 \;\; \textit{Volt - amperes}
\end{array}
$$

8400 VA – *lighting load*
3000 VA – *small appliance circuits-(220.52(A)*
<u>1500 VA – *laundry circuit* (20 *amperes*)</u>-(220.52(B)

12,900VA-Net Total

Range	12,000 VA	(100% of nameplate)
		(Table 220.55 not applicable)
Water heater	2500 VA	
Dishwasher	1500 VA	
Clothes dryer	5500 VA	
	34,400 VA Total	

220.82(B)

Demand Factor

First 10 kVA – 100% – 10,000 VA
Reminder @ 40%   –   9,760 VA (.40 × 24,400)
                              19,760 VA

220.82(C)

19,760 VA
 4,000 VA (40%of electric space heating for 4 or more units) (220.82(C)(5))
Total   23,760 VA

Note: AC Load-10 amperes × 240 volts = 2400 VA-this is less than the electric heating load, and may be omitted from the calculation

$$\frac{23,760\,VA}{240\,Volts} = 99\,amperes$$

$$Service\,Rating - \quad 100\,amperes\,(230.42 - 230.79(C))$$

$$\frac{\times .83\,(310.15(B)(7)(1))}{83\,amperes}$$

Table 310.15(B)(16)

3 AWG – copper –60°C-85 amperes-110.14(C)(1),(a)
1 AWG – aluminum –60°C-85 amperes-110.14(C)(1),(a)

Optional Calculation

Neutral Conductor

2800 square feet at 3 VA sq. it.	–	8400 VA
2 – Small appliance circuits	–	3000 VA
1 – Laundry circuit	–	1500 VA
		12,900 VA

First 3000 VA @ 100%   3000 VA
Remainder@ 35%   (12,900-3000VA ×.35)   3465 VA
Range – 8kVA   (220.55)@ 70% (220.61(B)(1))   5600 VA
Dryer – 5.5 kVA @ 70% (220.61(B)(1))   3850 VA
                              Dishwasher   1500 VA
                                                                         17,415 VA

$$\frac{17,415\,VA}{240\,V.} = 72.56\,or\,73\,amperes\ (220.5(B))$$

3 AWG-copper-85 amperes-60°C.-110.14(C)(1),(a)(1)
2 AWG-alumunum-75 amperes-60°C.-110.14(C)(1),(a)(1)

Multifamily Dwelling

Service – 240/120 volt -1-phase (20 dwelling units)
Each dwelling unit – 900 square feet
Each dwelling unit is equipped with a 10 kw range
Each dwelling unit has a washer (1500VA–220.52(B))
Each dwelling unit has an electric dryer   = 5000VA (220.54)
General lighting – 900 sq. ft. x 3 VA/ft^2 = 2700 VA (220.42)
Range   = 8000 VA (220.55)
Washer   = 1500 VA (220.52)(B))
Dryer   − 5000 VA (220.54)
Feeder size for each unit
General lighting -   2700 VA (220.42)
Small appliance load   3000 VA (220.52(A))
Laundry Load   1500 VA
                                                         7200 VA

First 3000 VA @ 100%   3000 VA
Remainder@ 35%   1470 VA (4200 VA × .35)
                                                        4470 VA Net (220.42)
Range Load   8000 VA (Table 220.55)
Dryer   5000 VA (220.54)
                                                       17,470 VA Total

$$\frac{17,470\,VA}{240\,Volts} = 73\,amperes\ -\ 80\,ampere\ feeder\ overcurrent\ device$$

Section 310.15(B)(7) – For one-family dwellings and the individual dwelling units of two-family and multifamily dwellings, the service and feeder conductors supplied by a single-phase, 120/240 volt system, or single-phase feeder conductors, consisting of 2 ungrounded conductors and the neutral conductor from a 208/120 volt, 3-phase, 4-wire, Wye connected system are permitted to have an ampacity of not less than 83% of the service or feeder rating. In this example, that is 83% of 80 amperes, or 66 (220.5(B)) amperes (310.15(B)(7)(1)(2)(3)), (215.2(A)(1).

Ungrounded feeder conductors – 4 AWG copper – 70 amperes - 60°C
(2 AWG aluminum -75 amperes 60°C.)

Neutral conductor – 4470 *VA* (220.42)
Range – 5600 *VA* (70% – 220.61(*B*)(1))
Dryer – 3500 *VA* (70% – 220.61(*B*)(1))
_____
13,570 *VA*

$$\frac{13,570}{240V.} = 57\,amperes$$

Neutral conductor   – 4 AWG – copper – 70° amperes – 60°C
– 3 AWG – aluminum – 65 amperes – 60°C

This wire size complies with 310.15(B)(7)(4)), as it meets the provisions of 220.61(B)(1).

Total Load for 20 Individual Dwelling Units

Lighting, small appliance, and laundry load – 20 units × 7200 VA = 144,000VA

Demand Factor (220.42)

First 3000 VA @ 100%   – 3,000 VA
Remainder at 35 %   – 49.350 VA (141,00VA @35%)
_____
52,350 VA

52,350 *VA* (General lighting and small appliance load)
35,000 *VA* (20 – 10 kw ranges) (Table 220.55- Column C)
_____
87,350 *VA*

Dryer load – 20 Dryers – 5000 VA, each
Table 220.54 – 42,770 VA

$$\begin{array}{r} 5000\,VA \\ \times\ \ 20\,Dryers \\ \hline 100,000\,VA \end{array}$$

20 dryers – 47 % - minus 1 % for each dryer above 11

$$\begin{array}{r} 100,000\,VA \\ \times\qquad .47 \\ \hline 47,000\,VA \end{array}$$

$$\begin{array}{r} \times\ \ .91\quad less\,9\%\,for\,9\,dryers) \\ \hline 42,770\,VA \end{array}$$

$$\begin{array}{r} 87,350\,VA \\ +42,770\,VA \\ \hline 130,120\,VA \end{array}$$

$$\frac{130,120\,VA}{240} = 542\ amperes$$

Feeder Conductors

542 amperes –2-300kcmil-THHN-copper –640 amperes@90°C., and 570 amperes @ 75°C. – 110.14(C)(1)(b)(1)

Neutral Conductor

52,350 VA – General lighting, small appliance load, and laundry load
24,500 VA – 35,000 VA × .70 (220.61(B)(1) – Range load
<u>29,939 VA</u> – 42,770 VA × .70 (Table 220.54) Dryer load (220.61(B)(1))
106,789 VA

$$\frac{106,789\,VA}{240} = 445\,amperes$$

200 amperes @100%
245 amperes @70%=171.5, or 172 amperes (*Demand Factor- 220.61(B)(2)*)

$$\begin{array}{r} 200\,amperes \\ +172\,amperes \\ \hline 372\,amperes \end{array}$$

Neutral conductor size – 500 kcmil – copper – 380 amperes - 75°C.

700 kcmil – aluminum – 375 amperes - 75°C.

Store Building

Service – Single-Phase – 240/120 Volts

Store – 80 feet by 90 feet = 7200 square feet –

– 50 feet of show window

– 90 – duplex receptacles

Table 220.12 – General Lighting Load by Occupancy

Stores – 3 volt – amperes per square foot

– 33 Volt – amperes per square meter

$$\begin{array}{r} 7200\,square\,feet \\ \times \quad 3\,VA\,(220.12) \\ \hline 21,600\,Volt\,-\,Amperes\,(General\,Lighting) \end{array}$$

Lighting load demand factor – 100 % - Table 220.42

Show–window lighting–200 volt–amperes per linear foot, or 300 mm (220.14(G))

$$\begin{array}{r} 200\,VA \\ \times \quad 50\,feet \\ \hline 10,000\,VA \end{array}$$

90–Duplex receptacles @ 180 volt – Amperes, each (220.14(I))

$$\begin{array}{r} 90\,-\quad Duplex\,receptacles \\ \times\,180\,\,VA \\ \hline 16,200\,\,VA \end{array}$$

Demand Factor (Table 220.44)

First 10,000 VA @100%

Remainder @ 50%

$$\begin{array}{r} 10,000\,VA \\ 3,100\,VA\,\,(50\%\,of\,6200\,VA) \\ \hline 13,100\,VA \end{array}$$

Outside sign circuit – 1200 volt–Amperes – (220.14)(F))

$$\begin{array}{l} 21,600 \quad Volt \quad Amperes \quad General\,Lighting \\ \underline{\times \quad 1.25 \quad (continuous\,load)} \\ 27,000 \quad Volt - Amperes \end{array}$$

$$\begin{array}{l} 10,000 \; Volt - amperes - Show\,Window\,Lighting \\ \underline{\times \; 1.25 \,(continuous\; load)} \\ 12,500 \; Volt - Amperes \end{array}$$

13,100 volt – amperes – Receptacle outlets (noncontinuous)

$$\begin{array}{l} 1,200 \; Volt\,Amperes - Outside\,sign\,circuit \\ \underline{\times 1.25 \,(continuous\; load)} \\ 1,500 \; Volt\,Amperes \end{array}$$

Continuous (Lighting) Load

$$\begin{array}{l} 27,000\,Volt - Amperes \\ 12,500\,Volt - Amperes \\ \underline{1,500\,Volt - Amperes} \\ 41,000 \; Volt - Amperes \end{array}$$

$$\begin{array}{ll} 41,000\,Volt - Amperes & (continuous) \\ \underline{13,100\,Volt - Amperes} & (noncontinuous)\,(receptacle\,outlets) \\ 54,100\,Volt - Amperes \end{array}$$

$$\frac{54,100\,Volt - Amperes}{240\,Volts} = 225\,Amperes$$

Service overcurrent device – 225 amperes (240.6(A))
Service conductors – 4/0 AWG – copper – 230 Amperes - 75°C.
　　　　　　　　110.14(C)(1),(b),(1)
　　　　　　　　300 kcmil – aluminum – 230 Amperes - 75°C.
　　　　　　　　110.14(C)(1),(b),(1)

*Service Conductors-

These are the conductors (or cable assembly) which extend from the **Service Point** to the service disconnecting means (Parts II and III, Article 230).

*Service Equipment-

This equipment is supplied by the Service Conductors, and from which the distribution system (feeders, branch-circuits) originates (Part VI, Article 230).

*Service Point-

This 'point' is the physical connection between the facilities of the serving utility and the premises wiring. It may be at the secondary terminals of a utility transformer, or at the line terminals of a circuit breaker or disconnecting means. This constitutes the beginning of the premises wiring system and the application of the NEC.

*Short-Circuit Current Rating-

There is a direct connection of this term to Section 110.10, which applies to Circuit Impedance, Short-Circuit Current Ratings, and Other Characteristics, and Section 240.4, and its Informational Note that refers to ICEA-P-32-382-2007- 'Allowable Short-Circuit Currents for Insulated Copper and Aluminum Conductors'.

And Section 110.24(A) discusses a requirement that 'Service Equipment', at other than Dwelling Units, be marked in the field with the 'Maximum Available Fault Current'. This requirement correlates with NFPA 70E - 2012, which provides guidance in determining exposure to fault conditions.

The following Sections apply to the short-circuit current rating of equipment:

1. 110.24(A)-Service Equipment
2. 230.82(3)-Meter Disconnect Switches up to 1000 Volts
3. 285.7- Surge Protective Device
4. 430.130(A)-Power Conversion Equipment
5. 440.4(B)- 440.10 HVAC Equipment
6. 646.7- Modular Data Centers
7. 670.5- Industrial Machinery

8. 700.5(E)-Transfer Switches (Emergency systems)
9. 701.5(D) -Transfer Switches (Legally-required standby systems)
10. 712.72- Microgrid Systems

In addition, Section 110.9 must also be considered in calculating the maximum available fault-current at the line terminals of equipment, so that the overcurrent devices will have the appropriate interrupting rating (capacity) to safely interrupt the fault current.

Also, HVAC equipment is required to have it's <u>short-circuit current rating</u> marked on the equipment nameplate (440.4(B)),(UL 1995). This does not apply to one and two family dwelling occupancies, cord-and-plug connected equipment, or equipment installed on a branch-circuit where the circuit rating does not exceed 60 amperes.

UL 1995 states that the HVAC equipment nameplate may specify the type of overcurrent device that must be used. If the nameplate specifies a 'Maximum Size of Overcurrent Device', either a fuse or circuit breaker may be used. If the nameplate specifies a 'Maximum Size Circuit Breaker', then a circuit breaker must be used. These are listing instructions and they must be followed to comply with 110.3(B).

The following is an example of calculating fault current from a transformer secondary to equipment downstream from the source.

> Transformer - 500 Kva
> Voltage (Secondary) - 480 v. - 3-Phase
> % Impedance (Nameplate) -1.3%
> Negligible Motor Load

Step 1. Full Load Secondary Current = $\dfrac{500{,}000 \text{ Va}}{480 \text{ V.} \times 1.732}$ = 601.42, or 601 amperes

Step 2. Establish Multiplier - $\dfrac{100}{1.3 \times 0.9 = 1.17}$ = 85.47

Note- % impedance (Z) may be + or − 10% − UL1561 - for transformers 25 kVA and larger.

Therefore, the worst case condition is minus 10% of transformer marked impedance.

Transformers built to ANSI Standards have a + or − 7.5% tolerance.
We will use -10% in this example.

$$
\begin{array}{r}
601 \quad amperes \\
\times \ \ 85.47 \quad (multiplier) \\
\hline
51{,}367\,A \quad Total\,Short - Circuit\,Current
\end{array}
$$

Available short-circuit current at the transformer secondary terminals is 51,367 amperes

Now extend a set of service conductors a distance of 50 feet (15.24 meters) to the location of the service equipment, consisting of 2-350 kcmil, copper conductors, per phase, in rigid metal conduit (310 amperes × 2 = 620 amperes, per phase) (Table 310.15(B)(16)).

Step 3. Identify the 'C' values for Conductors
　　　　　Table 4. – Eaton Selecting Protective Devices Handbook
　　　　　'C' value – 350 kcmil – copper – steel conduit = 19,704

Step 4. $\dfrac{1.732 \times 50 \text{ (feet)} \times 51{,}367}{19{,}704 \times 2 \text{ (conductors per phase)} \times 480 \text{ volts}}$ (fault current at transformer secondary)

　　　$= \dfrac{4{,}448{,}382}{18{,}915{,}840} = 0.2351$

Step 5. Establish Multiplier $- \dfrac{1}{1 + 0.2351} = 0.8096$

Step 6.　　51,367 amperes (at transformer secondary terminals)
　　　　　$\underline{\times\, 0.8096}$ (Multiplier)
　　　　　41,587 amperes available at the service equipment

This fault current would be marked on the service equipment to satisfy Section 110.24(A). This 'field marking' would also identify the date that the calculation was made. Also, please note that there is no motor contribution in this example.

To factor a <u>significant</u> motor contribution, multiply the transformer secondary full-load current rating by a factor of 4.

In our example, the full-load secondary current rating is 601 amperes.

601　*amperes*

$\underline{\times\ 4}$

2,404　*amperes*

Add this value to the calculated secondary short-circuit current to get the total short-circuit current at the transformer secondary.

　　51,367 amperes
　　$\underline{+\, 2{,}404}$ amperes (Motor Contribution)
　　53,771 amperes (Total Available Fault-Current at Transformer Secondary)

If the motor load represented only 50% of the total load, then the calculation would be:

$$
\begin{array}{r}
601 \text{ amperes} \\
\times \quad 4 \text{ (Motor Contribution)} \\
\hline
2404 \text{ amperes} \\
\times .5 \text{ (50\%)} \\
\hline
1202 \text{ amperes}
\end{array}
$$

After determining the available fault-current at the service equipment, this calculation method may be used to calculate the short-circuit current downstream, even to the farthest outlet supplied on a branch-circuit.

On another note, the 'C' values from the Eaton SPD Handbook for the same 350 kcmil copper conductors in <u>nonmetallic raceways</u> at 600 volts would be 22,737, instead of 19,704, where these conductors are installed in steel conduits.

Let's see how this effects the available short-circuit current at the service equipment.

$$
\frac{1.732 \times 50 \left(\text{feet}\right) \times 51,367}{22,737 \times 2 \text{ conductors per phase} \times 480 \text{ volts}} = \frac{4,448,382}{21,827,520} = 0.2037
$$

$$
\text{Establish Multiplier} = \frac{1}{1 + 0.2037} = 0.8307
$$

$$
\begin{array}{r}
51,367 \text{ amperes (at transformer secondary terminals)} \\
\times 0.8307 \\
\hline
42,671 \text{ amperes available at the service equipment}
\end{array}
$$

$$
\begin{array}{r}
42,671 \text{ amperes (nonmetallic conduit)} \\
- 41,587 \text{ amperes (steel conduit)} \\
\hline
1084 \text{ amperes (difference)}
\end{array}
$$

The magnetic properties of the steel conduit would reduce the available short-circuit current at the service equipment by 1084 amperes, due to the effects of hysteresis associated with the steel raceway.

*Simple Apparatus (as applied to Hazardous (Classified) Locations-

An electrical component or combination of components of simple construction with well defined electrical parameters that does not generate more than 1.5 volts, 100 mA, and 25 mW, or a passive component that does not dissipate more than 1.3 watts, and is compatible with the intrinsic safety of the circuit in which it is used.

Semiconductor devices, such as LEDs, inductors and capacitors that store and release energy and are within the power limits of this definition, and photocells, as well as thermocouples, that do not generate more than 1.5 volts, 100 mA, and 25mW, are examples of the simple apparatus defined here.

*Surge Arrester - Article 280-

*Surge Protective Device - Article 285-

I have combined these two terms because their purpose is similar and they serve the purpose of reducing the effects of transient voltages by limiting the effects of surge current.

A Surge Arrestor may be installed ahead of a supply transformer so as to discharge the surge energy associated with lightning induced influences from utility transmission lines.

These transmission lines act as a large antenna in the presence of electrically-charged thunder clouds. Electrical charges are predominately direct current, although there is an alternating current component, as well. These charges accumulate when warmer air from the earth surface rises and meets cooler air at higher elevations. This mixing of warmer and cooler air creates turbulence, which causes the rapid contact and separation of materials in the air (dust, dry ice, snow, etc.). The contact and separation of these materials causes static electricity to be produced within the cloud, and pockets of negative and positive charges begin to form. As the cloud mass moves through the air, opposite charges are induced on the earth. And these earthbound charges follow the cloud.

A passive lightning prevention system recognizes that the function of the strike termination devices of a typical lightning protection system are not to attract lightning from the cloud, but, to discharge the opposite polarity earthbound charges at the points of the strike termination devices, thereby removing the difference of potential between a building or structure and the opposite charges in the cloud mass.

An active lightning prevention system recognizes that the sharp points of the strike termination devices form a corona cloud above them that makes these prevention devices less attractive to the lightning strike, thus preventing the strike from damaging the building or structure.

An active lightning attraction system is designed to attract the lightning strike. This concept is thought to intercept the lightning down stroke (stepped leader) through the ionized air above the strike termination devices and provide a more attractive path for the lightning energy to flow to ground (electrode).

Lightning can occur within the cloud, or from cloud to cloud, or from the cloud to the earth.

As the positive and negative charges build within the cloud, the magnetic field associated with this charged mass begins to gain strength, and the air surrounding the cloud begins to ionize. The ionized air begins to lose its dielectric properties and the charges are free to flow toward the earth through a series of strokes, each about 150 feet (45.72 meters) in length. This is known as a stepped leader, and it creates a path that becomes closer to objects on the earth. Especially at higher elevations, or at isolated buildings or structures. It is at this time that the opposite earthbound charges may rise to meet the downward 'stepped leader', and a return stroke occurs. This is the bright streamer that we see, and it flows from the earth to the cloud through the conducting path that was formed by the 'stepped leader'. This process may develop several times before the charges in the cloud and their opposite earthbound charges are neutralized.

Many years ago, I found information regarding thunderstorm activity in the U.S. and Canada. This was an 'isokeraunic map', and areas were categorized by thunderstorm day activity. The term 'isokeraunic' is derived from two Greek words, 'iso' meaning 'same', and 'keraunous' meaning 'thunderbolt.' A 'thunderstorm day' is identified as one where thunder is heard at least once. Of course, topography plays an important role in this determination. The corridor between Tampa and Orlando, Florida is known as 'Lightning Alley', with as many as 90 'thunderstorm days' per year. Other areas, such as Southern California, have very few 'thunderstorm days'. This is a good parameter in determining the need for lightning and surge protection.

It is easy to see that exposed transmission and other power lines are a constant threat, due to the 'antenna effect' that they provide during thunderstorm activity. And the concern is not only due to a direct lightning strike, but also, of induced charges, that may affect these exposed conductors. A high voltage transient, and resultant surge current, could easily flow through these exposed lines. And the 'voltage wave' may cause surge current to flow in both directions through these conducting paths. Add to this the fact that where a power line terminates, the voltage wave can double in magnitude. Any electrical service near the termination of this power line would need additional protection. This may be a lightning protection system for the building or structure (NFPA 780), with additional protection in the form of a 'Surge Arrester', and one, or more, Surge Protective Devices downstream.

The following is pertinent information on this protection from Article 280 and Article 285.

*Surge Arrester – ANSI/IEEE C62.11-

This device is connected to each ungrounded conductor. Its rating must be equal to or greater than the maximum operating voltage at the point of application. For solidly grounded systems, the selection of the surge arrester is based on the phase-to-ground voltage.

If the system is ungrounded or impedance grounded, the selection is based on the phase-to-phase voltage.

Section 280.3 - This Section requires that the surge arrester be connected to each ungrounded conductor.

Section 280.12 requires that the connection to ground be made through a conductor that is no longer than necessary and is run in such a way as to avoid unnecessary, and especially sharp bends.

Section 280.21 states that this grounding conductor may be connected to the

1. grounded (neutral) service conductor
2. grounding electrode conductor
3. the service grounding electrode
4. equipment grounding terminal (ground bus) in the service equipment.

Section 280.23 requires that the connection between the surge arrester and the line conductors and the surge arrester and grounding connection be a minimum 6 AWG (13.30 mm) copper or aluminum conductor.

The reasoning behind the connection of the surge arrester and ground to be through a conductor that is 'no longer than necessary', and that this conductor 'avoid unnecessary bends', is to limit the impedance of this connection to ground. Once again, lightning currents are predominately 'Direct Current', but, there is an 'Alternating Current' component, as well. The frequency of this AC component is in the range of 3 kHz to 10 mHz. This frequency range is from moderate to very high, where the 'skin effect' of alternating current may be a problem, as it would be above 1 mHz. Excessive length of the grounding conductor, and unnecessary bends in the conducting path would increase its impedance. This could have a marked effect on increasing the rise in potential above the earth ('0' volts), with possible hazardous consequences for both people and equipment. It is for this reason that any connection to a grounding electrode (system), whether it be for a surge arrester, a surge protective device, a utility transformer, the service equipment grounding connection, or a separately-derived system grounding connection, be made through a conductor that is run as short and straight as possible (250.4(A)(1), Informational Note No.1).

Also, remember that transient voltages and associated surge currents are not only caused by lightning. The switching of large loads within a facility, as well as utility switching problems, such as circuit switchers within substations and switchyards that rapidly open and close, possibly several openings and closures in an attempt to clear lines of fault conditions, will also produce these effects which may affect downstream connections, even miles from the location of the fault.

It is imperative to locate the surge arrester close to the equipment to be protected. As a surge passes through the arrester, a wave will be reflected in both directions through the conductors that are connected to the arrester. The magnitude of this reflected wave increases as the distance from the arrester increases. Keeping the conductors as short as possible between the arrester and the equipment will significantly reduce this reflected wave and provide better protection. In this regard, the conductor connections from the surge arrester to line, bus, or equipment, and the connection to ground, will not be any longer than necessary. This provision is in recognition of the increased impedance of these conducting paths due to the increased impedance that is associated with the higher frequency of lightning (280.14).

*Article 285 - Surge Protective Devices (SPD) (1,000 Volts, or Less) - UL – 1449-

These devices, formerly known as Transient Voltage Surge Suppressors (TVSS), are classified as Type 1, Type 2, Type 3, or Type 4. These are 'listed' devices (UL-1449), and they may be installed on systems that are ungrounded, solidly grounded, or impedance grounded.

They have a marked short-circuit current rating (285.7) which must not be exceeded. This marking is not applicable to receptacle-type SPD's. Once again, short-circuit calculations must be made to assure that the short-circuit current ratings of these devices is not exceeded. Typically, the SPD at the service equipment will be a Class 1 device. And in order to comply with Section 110.24(A), a short-circuit analysis would have determined the available fault-current at the service equipment (except for dwelling units), and the available short-circuit current would be field marked on the service equipment. Additional fault-current calculations would be made downstream from the service point and the proper short-circuit current rated SPD's would be installed there, as well.

Section 285.12 discusses the method of routing the grounding conductor. The grounding conductor is run as short and straight as possible in order to limit the impedance of this connection. In our discussion of Surge Arrester grounding conductors, and the effects of lightning currents on these grounding conductors, we said that there is an AC component to lightning. The frequency of the AC component is relatively high, possibly from 3 kHz to 10 mHz. Due to this high frequency component, the length of the grounding conductor, and limiting the number of bends in this connection will reduce its impedance and limit the surge-current associated with lightning.

Section 285.23 - Type 1 SPD's are usually installed on the line-side of the service disconnecting means (230.82(4)). However, they may be installed at other locations, such as at the source of a separately-derived system, or on the load side of the first overcurrent device supplied by a separately-derived system (285.23 (A)(2)),(285.24(C)).

The connection of the Type 1 SPD to ground may be through the grounded service conductor, the grounding electrode conductor, or the equipment ground bus in the service equipment. Once again, make this connection in such a way as to limit the conductor length and to avoid unnecessary bends for important impedance reduction.

Type 2 SPD's are to be connected on the load side of the service disconnecting means (285.24(A)).

Type 2 SPD's may be connected on the line (supply) side of the service disconnect where installed as a part of listed equipment and the SPD is protected by properly sized overcurrent devices and there is a disconnecting means (230.82(8)).

Or, where a building or structure is supplied by a 'feeder', the Type 2 SPD may be connected anywhere, from the load side of the building or structure first overcurrent device, or elsewhere downstream (285.24(B)).

Or, for separately-derived systems, this device is to be connected on the load side of the first overcurrent device supplied by the separately-derived system (285.24(C)).

Type 3 SPD's are supplied by branch-circuits, and, they may be installed anywhere on the load side of a branch-circuit overcurrent device, up to the point where the branch-circuit conductors connect to the equipment. If specified by the manufacturer, the Type 3 SPD connection must be a minimum of 30 feet (10m) of conductor length from the service or separately-derived system disconnect (285.25).

SPD line and ground conductors must be a minimum of 14 AWG copper or 12 AWG aluminum (285.26).

Type 4 SPD's are sometimes provided by equipment manufacturers as part of 'listed' equipment. These devices are not 'field installed' (285.13).

*Switch, Bypass Isolation-

This switch is meant to 'bypass' equipment, usually for the purpose of maintaining a power supply to all, or part, of a facility while isolating one, or more pieces of equipment for the purpose of maintenance or repair (700.5(B), (701.5(B)).

*Switch, Isolating-

This switch is meant to isolate all, or part of a supply system. It is common on medium and high voltage systems (225.51),(230.204). This switch has no interrupting rating and would only be operated when the electrical load is removed.

The switch is provided with a means to connect the load side conductors to a grounding electrode when the switch is in the open position (230.204(D)).

Other applications for isolating switches include motors of over 100 HP, AC, or 40 HP, DC (430.109(E)). Once again, these switches have no recognized interrupting rating, and a sign must be placed at the switch indicating a warning –'Do not operate under load' (110.21(B)).

Capacitors or capacitor banks used on systems of over 1000 volts must have a means of isolating the capacitors from all sources of voltage. An isolating switch used for this purpose must have a sign stating the warning 'Do not operate under load'. (460.24(B)(1)(2)), (490.22), (110.21(B)).

In Class I, Division 2 Hazardous (Classified) Locations, isolating switches that serve transformers or capacitor banks that are not used as a means for disconnecting load current, may be installed in general-purpose enclosures (501.115(B)(2)).

*Thermal Protector (as applied to motors)-

This device, sensitive to heat, is an integral part of the motor and serves to disconnect the motor when excessive heat is detected in the motor winding. This device may be connected to a control circuit which would serve to disconnect the power circuit to the motor.

Section 430.32(A)(2) sets the limits for the trip-current of the thermal protector as follows:

Motor full-load current 9 amperes or less - 170%
Motor full-load current from 9.1 to, and including, 20 amperes - 156%

Motor full-load current greater than 20 amperes - 140%

Section 430.32(B) specifically addresses the use of a 'thermal protector' which is integral with the motor. This device consists of a set of normally-closed contacts that are attached to a bimetallic disk. Excessive heat in the heating coil of the thermal protector causes the bimetallic disk to expand and open the normally-closed contacts and stop the motor. A manual reset button will close these contacts after the motor has sufficiently cooled.

Most small motors of less than 1/20hp are 'impedance-protected', which means that the impedance of the motor windings will prevent dangerous overheating (430.32(B)(4)).

*Thermally Protected (as applied to motors)-

This reference applies to motors which have the words 'thermally protected' on the motor nameplate, indicating that the motor is inherently protected from damage due to overheating (430.32(A),(B)).

Where a motor is not inherently protected from overloads and separate overload protection is provided, 430.37 applies. These overload devices have no short-circuit interrupting rating and this protection is provided by the motor overcurrent protection. Section 430.32(A)(1) specifies that motors with a marked service factor of 1.15, or greater, are to be protected at no more than 125% of the motor nameplate current rating. A motor with a marked temperature rise not greater than 40° C. is to be protected at no more than 125% of the motor nameplate current rating. All other motors are to be protected at no more than 115% of the motor nameplate current rating.

However, Section 430.32(C) permits these percentages to be increased to 140% or 130% respectively, where the lower values are not sufficient for the starting or running current of the motor.

Section 430.36 recognizes that fuses may be used to provide overload protection, such as time-delay fuses. Inverse-time circuit breakers will not operate at 115% or 125% of the motor nameplate current in a short enough time (or at all) to protect the motor from an overload.

Table 430.37 identifies the 'overload units' that are required for single-phase, two-phase, and three-phase motors.

It is interesting to note that before 1971, three-phase motors required only two overload units, typically in Phases A and C. Phase B was unprotected. This led to problems of motor damage due to 'single phasing'. Whether the supply system is Wye-Delta or Delta-Wye, if a primary phase becomes open, the 3 secondary phases will still be supplying current to the motor. However, with only 2 overload units in Phases A and C, the current in these 2 phases may be 115% (1.15) of the normal current. But, in the unprotected Phase (B), the current may be 230% (2.3) of the normal current. This single-phasing condition may, and probably will, cause damage to the motor. In recognition of this problem, the 1971 edition of the NEC was modified to require overload protection in each phase of a three-phase motor.

*Utilization Equipment-

Equipment that utilizes electric energy for electronic, electromechanical, chemical, heating, lighting, or similar purposes.

The following references apply to supply circuits and equipment associated with utilization equipment.

Section 210.23 covers 'Permissible Loads' on branch circuits. The total load may not exceed the rating of the branch circuit.

Section 210.23(A)(2) specifies that the total load of utilization equipment, other than luminaires, that is fastened in place (fixed), is not to exceed 50% of the branch-circuit ampere rating where lighting units, and, or, cord-and-plug connected utilization equipment is also supplied.

A 30, 40, or 50 ampere branch-circuit may be used to supply utilization equipment. In dwelling occupancies, a 30 ampere branch circuit may supply a cord-and-plug connected utilization equipment (e.g., electric clothes dryer), where this equipment does not exceed 80% of the branch circuit rating. Also, a 30 ampere branch circuit may supply lighting units with heavy-duty lamp holders in other than dwelling occupancies, or utilization equipment in any occupancy (210.23(B)).

Section 210.23(C) permits 40 or 50 ampere branch circuits to supply utilization equipment (cooking appliances) in any occupancy.

Section 210.23(D) permits circuits larger than 50 amperes to supply utilization equipment. This is in conjunction with Section 210.3, Exception, where these circuits supply industrial equipment.

Section 314.24 covers the topic of the depth of outlet or device boxes. This would include boxes that are used for the connection of utilization equipment.

Section 314.27(D) addresses boxes that may be used to support utilization equipment, other than ceiling supported fans.

Section 513.10(E)(2) addresses the use of portable utilization equipment in an aircraft hangar. This equipment must be of a type that is suitable for Class I, Division 2 or Zone 2 locations. Flexible cords for the portable equipment must be of the extra hard usage type (Table 400.4), and the flexible cord must have a separate equipment grounding conductor.

Section 515.7(C) applies to portable luminaires and utilization equipment installed above a Hazardous (Classified) Location.

Section 516.4(D) prohibits portable luminaires and <u>utilization equipment</u> in the spraying area of a spray booth. However, Exception No. 1 recognizes portable luminaires that are identified for Class I, Division 1, or Class I, Zone 1 locations as suitable for this environment.

Exception No. 2 recognizes portable electric drying equipment within a spray booth. This drying equipment, within 18 inches (450 mm) of the floor, must be suitable for Class I, Division 2, or Class I, Zone 2 locations. And this equipment is not to be in use during spraying operations, and the equipment has all metallic parts properly bonded and grounded.

Also, interlocks are provided to prevent the operation of the spraying equipment while the drying equipment is within the spray booth. In addition, there must be a 3-minute purge of the spray booth before energizing the drying equipment, and the drying equipment must remain inoperable on failure of the ventilation system.

*Volatile Flammable Liquid-

NFPA 30 - Flammable and Combustible Liquids Code-Flammable liquids are those that have 'flash points' below 100° F, or 38° C. The flash point of a liquid is the temperature at which the liquid produces an ignitable vapor above the surface of the liquid. Combustible liquids have flash points of 100° F., or 38° C., and higher.

The flash points of these liquids have lower and upper flammable limits which identify the ignitable range of concentration when mixed with air. Ignition of the vapor will occur within the range of the flammable limits. If ignition occurs near the lower or upper flammable limits, explosion pressures will be relatively weak. Maximum explosion pressures occur when the mixture is in the mid-range between the lower and upper flammable limits. Explosion pressures are also affected in accordance with the energy of the ignition source. That is, the higher the ignition energy, the greater the explosion pressures created.

Flammable liquids are identified into three classes.

Class IA - Flash Point below 73° F., and Boiling Point below 100° F.
    Examples include:
    ethyl chloride
    ethyl ether
    acetaldehyde

propylene oxide

Class IB - Flash Point below 73° F., and Boiling Point at, or above, 100° F.
Examples include:
acetone
benzene
gasoline
toluene

Class IC - Flash Point at, or above, 73° F., and Boiling Point at, and below 100° F.
Examples include:
isopropanol
methyl alcohol
turpentine
styrene

Combustible Liquids are identified into three classes.
Class II, with Flash Points at, or above 100° F., and below 140° F.
Class IIIA, with Flash Points at, or above 140° F., and below 200° F.
Class IIIB, with Flash Points at, or above 200° F.

Gases and vapors are divided into four groups (A, B, C, or D), depending on their explosion characteristics. This information is an aid in determining the suitable equipment for the hazardous environment (500.6) (Article 501).
Group A - Acetylene - Ignitable range - 2 1/2% to 81% in normal air.
Group B - Hydrogen - Ignitable range - 4% to 75% in normal air.
Group C - Ethylene - Ignitable range - 2.75% to 28.6 % in normal air.
Group D - Propane - Ignitable range - 2.1% to 10.1% in normal air.

Another important consideration is the Flammable Liquid vapor density. Or, how heavy or light the vapor may be, as compared to normal air. The vapors produced by styrene are very heavy, about 3.6 times the weight of normal air. If not contained, this heavy vapor may spread for great distances away from the source. This may have a significant effect on the extent of the Hazardous (Classified) Location, as compared with a lighter material, such as hydrogen, which is only about 10% of the weight of normal air.

In addition, all potential sources of ignition must be carefully considered. And this would be especially important in the presence of a heavier than air vapor, such as produced by styrene. These sources of ignition would include static electricity, which would normally have more than enough ignition capable energy.

In fact, a thermite reaction, caused by a ferrous tool striking a concrete floor, is certainly a cause for concern where heavier than air vapors are present.

*Voltage, of a Circuit-

This term is defined as the greatest rms difference of potential between any 2 conductors of the circuit.

*Voltage, Nominal-

This is the value of voltage that is typically used for load calculations. Section 220.5(A) identifies the nominal voltages for branch-circuit and feeder calculations. In addition, Informative Annex D states that for uniform application of Articles 210, 215, and 220, a nominal voltage of 120, 120/240, 240, 208Y/120 is used to calculate the ampere load on a conductor.

*Voltage to Ground-

Normally, this voltage would be between the ungrounded conductor(s) and the conductor that is intentionally grounded. For ungrounded systems, the voltage between the ungrounded conductors.

*Watertight-

Constructed so that moisture will not enter the enclosure under specified test conditions. Enclosures marked as Type 4, 4X, 6, or 6P, may also be identified as 'Watertight'. These enclosures are tested in a stream of water (110.28).

*Waterproof-

Constructed in such a way that weather conditions will not interfere with the operation of the contained equipment. Depending on certain conditions, enclosures identified as Types 3, 3S, 3R, 4, 4X, 6 or 6P may be considered as 'Waterproof' (Table 110.28).

However, I have seen many installations that are Hazardous (Classified) Locations, where 'Explosionproof' enclosures, with flanged cover joints, are improperly installed in wet locations. The flat cover joint has a purposeful gap of .0015″ (0.4mm) to allow the enclosure to breathe. This enclosure, which may have the appearance of being watertight, should not be installed in a wet location, unless it is also identified for this type of environment.

Over 1000 Volts, Nominal

*Electronically-Actuated Fuse-

These devices are equipped with a control module to sense current changes and initiate the opening of the fuse when an overcurrent occurs. In this way, they function much the same as an electronic circuit breaker. They do not rely on the heating of a fuse element, as in a typical fuse. Also, they do not provide protection against overloads, and they may be designed to provide for current limitation (current-limiting).

The trip setting of the control module should be designed so that selective coordination is achieved with downstream overcurrent devices.

*Fuse-

An overcurrent device with either a single fusible link or a multiple fusible link that is sensitive to the passage of current. It may be identified as 'current-limiting', in which case the clearing time is 1/2 cycle (.008 seconds), or less. In addition, it may have time-delay characteristics which would lend itself as desirable for certain types of loads.

For voltage of over 1000 volts, information relating to Power Fuses is found in Section 490.21(B).

*Expulsion Fuse Unit-

This is a vented fuse which allows the venting of gases that are produced by an internal arc. The opening of the fuse may be aided by a spring. These fuses are not current-limiting, so they do not limit the magnitude of the fault current.

*Controlled Vented Power Fuse-

This fuse is designed to control the release of any solid material into the surrounding atmosphere during the interruption of current. The discharge of gases during current interruption will not damage insulation or lead to a flashover between conductors, or between a conductor and a grounded member.

*Nonvented Power Fuses-

These fuses are designed to prevent the release of gases, liquids, or solid materials during the interruption of fault current.

*Vented Power Fuse-

This fuse is designed to release the energy of an arc in the form of gases, liquids, or solid material to the surrounding environment. This type of fuse would be restricted in Hazardous (Classified) Locations, due to the release of ignition-capable energy (501.115),(502.115). These fuses are not permitted to be installed indoors, underground, or in metal enclosures, unless identified for this purpose (490.21(B)(1),(5)).

*Circuit Breaker-

This device is described as one operating to switch, carry, and interrupt current under normal or fault conditions. At voltages above 600 volts, and, indeed, above 800,000 volts, a circuit breaker must be capable of opening under normal circuit conditions, as well as under short-circuit conditions. At higher voltages, the arcing produced by opening the circuit breaker contacts may be cooled and quenched by being within a vacuum, or immersed in oil, or, where the current-carrying contacts operate in sulfur hexafluoride (SF6). The closing rating of the circuit breaker must be at least equal to the maximum available asymmetrical fault current. And, the interrupting rating of the circuit breaker must be at least equal to the maximum fault current that the circuit breaker will be required to interrupt (490.21(A)(4),(2)).

The voltage rating of the circuit breaker must be at least equal to the circuit voltage (490.21(A)(4),(5).

Section 490.21(A)(1) covers the location of circuit breakers. Typically, they must be mounted in metal enclosures or within fire-resistant cells. However, open-mounted units are acceptable where accessible to only qualified persons.

If circuit breakers are installed to control oil-filled transformers within a transformer vault (Article 450, Part III), they must be installed outside the vault, or be capable of being operated from outside the vault (490.21(A)(1),(b)).

*Disconnecting (or Isolating) Switch-

Isolating switches may be used to deenergize equipment, typically for maintenance or repair. If these switches are not interlocked with devices that are designed as circuit-interrupting devices, there must be a warning sign to indicate that this switch is not to be opened under load (110.21(B), (490.22).

Part VIII of Article 230 applies to 'Services Exceeding 1000 volts, nominal'. Section 230.204 addresses the use of 'Isolating Switches'. These switches are normally required on the <u>line side</u> of the service disconnect, which would be an air-break switch, an oil, vacuum, or sulfur hexafluoride circuit breaker. The isolating switch, with visible break contacts, has no inherent interrupting capabilities, and must not be opened under load conditions. Section 110.21(B) would apply here to require appropriate visible and durable signage at the isolating switch location to indicate that the switch is not to be operated under load. Section 230.204(C), (225.51) requires that the isolating switch be made accessible to qualified persons only.

Section 230.204(D) requires a direct connection to a grounding electrode system, equipment ground busbar, or grounded structural steel, when the isolating switch is in the open position.

*Oil Filled Cutout-

This device has the fuse support, fuse link, and switch blades immersed in oil. The oil acts as an insulator and serves to quench any arcing associated with normal switching or arcing under fault conditions.

The interrupting rating and fault closing rating must be equal to, or greater than the available (asymmetrical) fault current at the cutout location (490.21 (D)(2),(4)).

The continuous current rating and voltage rating must be suitable for the maximum current through the cutout, and the maximum circuit voltage (490.21(D)(1),(3)).

*Regulator Bypass Switch-

This may be one or more switching devices to allow for the bypassing of a voltage regulator (490.23). If any of these switching devices are used, including circuit breakers, cutouts, and bypass switches, and there is a possibility of a back feed, a suitable warning sign must be provided to indicate this hazard (110.21(B)). Or, a permanent and legible one-line diagram must be provided within sight of each point of connection which would indicate the switching arrangement (490.25).

To Convert U.S. Customs Units of Measurement to Metric Sizes

Multiply the U.S. Measurement by 25.4, or
Divide the Metric Size by 25.4 to convert to U.S. Measurement

To convert circular mils to metric wire sizes, divide the circular mil area by 1973.53, or multiply the metric wire size by 1973.53 to convert to circular mils.

Inch	=	0.254 meters
Inch	=	2.54 centimeters
Inch	=	25.40 millimeters
Meter	=	39.37 inches
Millimeter	=	0.03937 inch
Centimeters	=	inches/2.54
Foot	=	.3048 meter
Yard	=	0.9144 meters
Mile	=	1609 meters
Kilometer	=	0.6213 miles
Square meter	=	square foot ÷0.093
Circumference	=	πd
Area of circle	=	πr^2
Celsius to Fahrenheit	=	temperature × 1.8 + 32
Fahrenheit to Celsius	=	temperature −32 × .5556
Square feet to square meter	=	$m^2 = \dfrac{ft^2}{10.764}$
π	=	3.1416
$\sqrt{2}$	=	1.414
$\sqrt{3}$	=	1.732
Voltage-drop	=	$\dfrac{2K \times L \times I}{CM}$ - single-phase
Voltage-drop	=	$\dfrac{1.732K \times L \times I}{CM}$ - three-phase
Circular mils	=	$\dfrac{2K \times L \times I}{VD}$ - single-phase
Circular mils	=	$\dfrac{1.732K \times L \times I}{VD}$ - three-phase

K = 12.90 ohms – copper
K = 21.20 ohms – aluminum
L = one way length in feet of conductor
I = amperes of load
CM = circular mil area of conductor (NEC Table 8-Chapter 9)
Neutral current in a 3-phase, 4-wire, Wye system

$$\sqrt{L1^2 + L2^2 + L3^2 - \left[(L1 \times L2) - (L2 \times L3) - (L1 \times L3) \right]}$$

Series Circuits

Total Resistance - $R_T = R_1 + R_2 + R_3 + R_4$
Resistance Where Voltage and Power (Wattage) are Known
$$R = E^2 P$$
Power (Wattage) – $P = I^2 \times R$
Voltage Equals the Sum of all of the Power Supplies
$$V_T = V_1 + V_2 + V_3 + V_4$$
Current Equals- $I = E$ (Source) / R (Total)

Parallel Circuits

The voltage – drop across each resistance is equal to the voltage of the power source.

The current in each branch of the parallel circuit is calculated by the formula - $I = \dfrac{E}{R}$

Parallel Circuit Power (Wattage) in each branch equals $P = I^2 R$ or $P = E \times I$ or $P = \dfrac{E^2}{R}$

Total Power - The sum of the power in each branch of the parallel circuit equals the total power consumed.

If all of the resistors have the same resistance, the total resistance equals the resistance of one resistor, divided by the total number of resistors in parallel.

Example

What is the total resistance of 5 – 10 ohm resistors in parallel?

$$\frac{10\,ohms}{5\,resistors} = 2\,ohms\,Total$$

Where the paralleled resistors have different ohmic ratings, the total resistance may be calculated using the Product Over Sum Method

Example

Two resistors in parallel – 5 ohms – 7 ohms

$5\,ohms$	$5\,ohms$
$\times 7\,ohms$	$+7\,ohms$
$35\,ohms$	$12\,ohms$

$$\frac{35\,ohms}{12\,ohms} = 2.92\,ohms$$

Total resistance = 2.92 ohms

If a third resistor is added (fourth, fifth, sixth, etc.), the same formula may be used to calculate the total resistance.

Example

Two resistors are added, one is 10 ohms, and the second is 15 ohms, what is the total resistance?

$$\begin{array}{r} 2.92\,ohms \\ \times 10\,ohms \\ \hline 29.2\,ohms \end{array} \qquad\qquad \begin{array}{r} 2.92\ ohms \\ +10\,ohms \\ \hline 12.92\ ohms \end{array}$$

$$\frac{29.2\ ohms}{12.92\ ohms} = 2.26\,ohms$$

$$\begin{array}{r} 2.26\,ohms \\ \times 15\,ohms \\ \hline 33.9\,ohms \end{array} \qquad\qquad \begin{array}{r} 2.26\ ohms \\ +15\,ohms \\ \hline 17.26\ ohms \end{array}$$

$$\frac{33.9\ ohms}{17.26\ ohms} = 1.96\,ohms$$

Total resistance = 1.96 ohms

The <u>Reciprocal Method</u> may be used to calculate the total resistance where the resistances have different ratings.

Example

Four resistors in parallel with ratings of 10 ohms – 6 ohms – 8 ohms – 15 ohms

$$\frac{1.00}{\dfrac{1}{10} + \dfrac{1}{6} + \dfrac{1}{8} + \dfrac{1}{15}}$$

$$\frac{1}{10} = 0.1 - \frac{1}{6} = 0.167 - \frac{1}{8} = 0.125 - \frac{1}{15} = 0.067$$

$$
\begin{array}{r}
0.100 \\
+0.167 \\
+0.125 \\
+0.067 \\
\hline
0.459
\end{array}
$$

$$\frac{1.00}{0.459} = 2.178\,ohms$$

Total resistance equals 2.178 ohms

The NEC and You Perfect Together

The following 200 questions, will serve to reinforce your understanding of the NEC, and if you are preparing to take an exam for an electrical license, this exercise is an invaluable aid in your exam preparation. Try to answer the questions first before reviewing the answer key, and grade yourself in the process. Be sure to carefully review the appropriate NEC Sections, Exceptions, and Informational Notes.

As always, I welcome your comments and suggestions.

Gregory P. Bierals
Electrical Design Institute
March 24, 2021

1. Liquidtight Flexible Nonmetallic Conduit must be bent in such a way as not to damage the raceway and not to reduce its diameter. Bends may be made
   A) By using a manual bender
   B) With equipment supplied by the manufacturer
   C) By hand shaping the conduit as necessary
   D) By using a bending machine

2. What type of equipment may not be installed over the steps of a stairway?
   A) Overcurrent devices
   B) Toggle switches
   C) Motor circuit switches
   D) Luminaires

3. The NEC permits fixture wire to be as small as
   A) 18 AWG
   B) 16 AWG
   C) 14 AWG
   D) 20 AWG

4. Class 2 and Class 3 wiring that is not terminated on equipment and not tagged for future use is considered abandoned.
   A) True
   B) False

5. Where a separate grounding electrode is provided for radio and TV equipment (Article 810), what is the minimum size bonding jumper required for connection to the building grounding electrode system?
   A) 6
   B) 8
   C) 12
   D) 10

6. There is no requirement to provide a grounding electrode at a separate building or structure that is supplied by a single branch circuit.
   A) True
   B) False

7. Where switchgear is installed indoors, what is the minimum height of dedicated space above the equipment enclosure, unless the structural ceiling is lower than this dimension?

A) 6 feet
B) 4 feet
C) 3 feet
D) 5 feet

8. Legally required standby systems are required to be tested and found to be acceptable to the AHJ when initially installed and no further tests are required.
   A) True
   B) False

9. UF cable must have the minimum cover requirements specified in 300.5(A). Where this cable is installed for a 120 volt-15 ampere GFCI protected circuit under the driveway of a one-family dwelling the minimum cover must be at least.
   A) 12 inches
   B) 24 inches
   C) 18 inches
   D) 6 inches

10. Which Article covers the installation and the requirements for generators?
   A) 450
   B) 490
   C) 445
   D) 430

11. Solidly grounded, 3-phase, 4-wire, Wye electrical systems supplying premises wiring require ground-fault protection where the system operates at over 150 volts-to-ground and up to 1000 volts phase-to-phase, where the service disconnecting means is rated at ____, or more.
   A) 1200 A
   B) 1000 A
   C) 3000 A
   D) 2000 A

12. What type of enclosure provides protection from exposure to the weather so that the successful operation of the internal components is not affected?
   A) raintight
   B) rainproof
   C) weathertight
   D) weatherproof

13. Where a pull or junction box has any dimension over 6 feet the conductors must be cabled or racked to avoid damage.
A) True
B) False

14. Where a feeder supplies _____ and equipment grounding conductors are required, the feeder conductors must include an equipment grounding conductor.
A) equipment
B) branch circuits
C) overcurrent devices
D) switchgear

15. Communications cables and coaxial cables for CATV systems are required to be separated from lightning down conductors by at least _____ feet, where practicable.
A) 3
B) 6
C) 2
D) 10

16. Patient care spaces in health care facilities are required to be supplied by a branch circuit that includes an insulated equipment grounding conductor, and the metal face plates for switches and receptacles are properly grounded by their mounting screws.
A) True
B) False

17. Back-fed circuit breakers may be of the plug-in type for special applications, such as a stand-alone inverter connected to a stand-alone PV system.
A) True
B) False

18. May ground-fault protection of equipment be provided for a fire pump circuit?
A) True
B) False

19. Bonding devices used for the grounding of the metal frames of PV modules and other equipment must be _____.
A) listed
B) labeled

C) identified
D) all of these

20. Ground-fault protection for a PV system must isolate the _____ conductors, or the inverter charge controller fed by the faulted circuits shall stop the supply to output circuits.
A) ungrounded
B) equipment grounding conductor
C) neutral
D) grounded

21. Explosionproof seals are required in Class II Hazardous (Classified) Locations.
A) True
B) False

22. Motors identified for Class I, Division 2 Hazardous (Classified) Locations are suitable for Class I, Division 1.
A) True
B) False

23. LFMC is required to be supported every 4½ feet and within 12 inches of the conduit termination, except where fished through _____ spaces in finished buildings or structures.
A) exposed
B) concealed
C) open
D) blocked

24. UF cable is not permitted for inside wiring.
A) True
B) False

25. NUCC is permitted to be used for inside wiring.
A) True
B) False

26. Galvanized steel EMT may be used in direct contact with the earth.
A) True
B) False

27. USE cable may be used for interior wiring.
    A) True
    B) False

28. Running threads are permitted for coupling connections on RMC and IMC.
    A) True
    B) False

29. Fittings and accessories (bolts, screws, straps, etc.) for RMC and IMC in wet locations shall be _____.
    A) protected by corrosion-resistant materials
    B) of corrosion resistant material
    C) weather-resistant
    D) A or B

30. Metallic enclosures are required to be _____ to comply with Article 250.
    A) bonded
    B) secured
    C) grounded
    D) A and C

31. Where equipment is supplied by Class 1 circuits operating at less than 50 volts, grounding is not required.
    A) True
    B) False

32. Type 2 surge protective devices may be connected on the line side of the service overcurrent device.
    A) True
    B) False

33. The conductor size for line and ground connections on a surge protective device shall be not smaller than 14 AWG copper or 12 AWG aluminum.
    A) True
    B) False

34. Surge protective devices shall be _____.
    A) listed
    B) identified

C) approved

D) A and B

35. Receptacle type surge protective devices are required to be marked with a short-circuit current rating.

A) True

B) False

36. Surge protective devices shall be connected to ground by a conductor that is not any _____ than necessary and shall avoid any _____ bends.

A) shorter

B) longer

C) unnecessary

D) B and C

37. Type 1 surge protective devices may be connected to the _____.

A) grounded service conductor

B) grounding electrode conductor

C) equipment grounding terminal in the service equipment

D) All of these

38. A surge arrester on an impedance or ungrounded system shall have a continuous operating voltage of the _____.

A) Phase-to-Phase voltage

B) Phase-to-Ground voltage

C) Phase-to-Phase voltage × 1.732

D) Phase-to-Neutral voltage

39. Intrinsically safe apparatus, including metallic raceways and enclosures, are not required to be grounded.

A) True

B) False

40. Where a building or structure is supplied by a feeder, a maximum of _____ switches or circuit breakers may serve as the disconnecting means.

A) six

B) four

C) two

D) one

41. Fuel cell systems shall be listed or _____ for the intended application.
    A) field labeled
    B) ratified
    C) approved
    D) recognized

42. Where circuits are rated over 100 amperes, the equipment terminal provisions are listed for use with _____ conductors.
    A) 60°C
    B) 75°C
    C) 90°C
    D) 105°C

43. Where service metal raceways and metal-clad cable assemblies are used with threadless fittings they must be _____ to be considered electrically continuous.
    A) secured
    B) weatherproof
    C) made up tight
    D) listed

44. Coaxial cables are permitted in a raceway with nonconductive and conductive optical fiber cables in compliance with Parts I and V of Article 770.
    A) True
    B) False

45. Branch circuit conductors shall be a minimum size of _____ AWG copper for voltages up to 2000 volts.
    A) 12
    B) 10
    C) 14
    D) 8

46. Where the AC ungrounded service entrance conductors are 3/0 AWG aluminum, a copper grounding electrode conductor may be _____.
    A) 8 AWG
    B) 6 AWG

C) 4 AWG

D) 2 AWG

47. For concrete encased electrodes, a copper grounding electrode conductor shall not be required to be larger than _____, unless this conductor is extended to other electrodes that require a larger conductor.

A) 6 AWG

B) 4 AWG

C) 3 AWG

D) 1 AWG

48. A grounding electrode for the service is not required to be bonded to a grounding electrode for the communications system if this electrode is more than 50 feet away.

A) True

B) False

49. If there are no service entrance conductors, the grounding electrode conductor shall be determined by the _____ size of the largest service entrance conductor required for the load to be served.

A) equivalent

B) neutral

C) grounded

D) 250.66

50. A - 5/8″ diameter copper-clad steel ground rod has the equivalent current-carrying capacity of a 3/0 AWG copper conductor (200 amperes @ 75°C.). If the single ground rod has a _____ ohm resistance to the earth, it does not have to be supplemented by another electrode.

A) 20 ohms

B) 5 ohms

C) 25 ohms

D) 10 ohms

51. The metal forming shell (structural reinforcing steel) of a swimming pool may be used as part of the grounding electrode system.

A) True

B) False

52. The frame of a portable generator _____ be required to be connected to a grounding electrode where the generator supplies only equipment mounted on the generator, or cord-and-plug equipment supplied from receptacles mounted on the generator.
    A) shall
    B) shall not
    C) may
    D) may not

53. Electrical systems that are grounded shall be connected to the earth in a manner that will limit the _____ imposed by lightning, line surges, or unintentional contact with higher voltage lines and _____ the voltage to earth during normal operation.
    A) amperage
    B) voltage
    C) stabilize
    D) B and C

54. The grounding electrode (system) for a lightning protection system _____ be bonded to the building or structure grounding electrode system.
    A) shall
    B) shall not
    C) is prohibited to
    D) may not be

55. Buildings or structures that support a photovoltaic system array are required to have a grounding electrode system in accordance with Article 250, Part III.
    A) True
    B) False

56. A system bonding jumper is associated with _____.
    A) a service supplied system
    B) a separately-derived system
    C) Table 250.102(C)(1)
    D) B and C

57. A solar photovoltaic system may be considered a separately-derived system.

A) True
B) False

58. Where the ungrounded conductors are increased in size to compensate for circuit conditions, such as voltage-drop, a _____ increase in size must be made to the equipment grounding conductor, unless the AHJ permits a qualified person to determine the size of the EGC.
   A) proportional
   B) raceway
   C) fuse
   D) circuit breaker

59. Equipment grounding conductors for photovoltaic source and output circuits shall be sized in accordance with 250.122. Increases in size to address voltage drop _____ be required.
   A) shall
   B) shall not
   C) insulated conductors
   D) A and C

60. Type AC cable has an outer covering that is an equipment grounding conductor.
   A) approved
   B) identified
   C) listed
   D) B and C

61. The feeder calculated load for 5-3kW household electric cooking appliances may be determined by Column _____ of Table 250.55.
   A) A
   B) B
   C) C
   D) A and C

62. An apartment complex with 50 individual units and provisions for an electric dryer in each unit may apply a demand factor of _____ to the total connected load.
   A) 50 %
   B) 60 %
   C) 25 %
   D) 100 %

63. Nondwelling receptacle loads, calculated at 180 volt-amperes each, may be calculated with no demand factor for the first 10,000 volt-amperes, and the remainder at _____.
    A) 50 %
    B) 70 %
    C) 50 %
    D) 40 %

64. If rock bottom is encountered and driving a ground rod at up to 45 degrees is not possible, the rod may be buried horizontally in a trench that is _____ deep.
    A) 1 feet
    B) 2 feet
    C) 2½ feet
    D) 3 feet

65. If locating a motor disconnect switch within sight from the motor controller increases the hazard to people or property, such as in a Hazardous (Classified) Location, the disconnect may be remotely located.
    A) True
    B) False

66. On a 3-phase, 4-wire, Wye connected system, the common point is defined as the neutral point.
    A) True
    B) False

67. The NEC contains provisions that are adequate for good service or future expansion of electrical use.
    A) True
    B) False

68. Nonmetallic boxes may be used with
    A) nonmetallic sheathed cables
    B) nonmetallic conduits
    C) metallic raceways
    D) A and B

69. Wireless power transfer equipment for electric vehicles is considered a _____.
    A) portable power system

B) separately-derived system
C) personnel protection system
D) electric vehicle coupler

70. What size copper grounding electrode conductor is required for a separately derived system that has two sets of 500 kcmil aluminum conductors, per phase?
A) 2/0
B) 4
C) 6
D) 1/0

71. A grounding electrode for strike termination devices (lightning protection) must be not less than _____ feet from any other electrode of another grounding system.
A) 5 feet
B) 10 feet
C) 6 feet
D) 8 feet

72. Multiple ground rods should be spaced at least twice the length of the longest rod for optimum paralleling efficiency.
A) True
B) False

73. Where electric vehicle supply equipment is _____for charging electric vehicles indoors without ventilation, no ventilation is required.
A) approved
B) certified
C) listed
D) identified

74. If an auxiliary grounding electrode is used to supplement an equipment grounding conductor specified in 250.118, the _____ requirements of 250.50, 250.53(C), or 250.58 do not apply.
A) grounding
B) bonding
C) supplemental
D) all of these

75. A _____ controls another circuit through a relay or similar device.
    A) signal circuit
    B) secondary circuit
    C) annunciator
    D) remote – control circuit

76. A _____ rating must be marked on a surge protective device.
    A) ampere
    B) watts
    C) short-circuit current
    D) all of these

77. If multiconductor cables are installed in parallel in the same cable tray a _____ equipment grounding conductor may be used in combination with the equipment grounding conductors within the multicondutor cable assemblies.
    A) copper
    B) single
    C) larger
    D) smaller

78. Type MC cable, listed as an equipment grounding conductor, or supplemented with an internal insulated equipment grounding conductor, is acceptable in an Assembly Occupancy.
    A) True
    B) False

79. Automatic transfer switches shall be _____ for emergency system use.
    A) listed
    B) approved
    C) identified
    D) A and C

80. Automatic transfer switches for emergency systems must be field marked to indicated the _____.
    A) use
    B) short-circuit current rating

C) location

D) load shedding

81. The allowable ampacity of 16 AWG fixture wire is _____.

A) 8 A

B) 6 A

C) 10 A

D) 17 A

82. Where conductors are spliced or tapped in a metal wireway, the maximum fill at the splice or tap cannot exceed _____.

A) 20%

B) 75%

C) 40%

D) 30%

83. It is a requirement to extend the solid 8 AWG copper bonding conductor for equipotential bonding of a pool or spa to the service equipment.

A) True

B) False

84. _____ shall be installed to give notice of electric shock hazard risk to persons swimming near a boat dock or marina.

A) Temporary signs

B) Permanent safety signs

C) Recreational swimming only signs

D) Swimming at night only signs

85. Power systems that are derived within listed Information Technology Equipment that are supplied as part of this equipment are not considered to be separately derived in accordance with 250.30.

A) True

B) False

86. The wiring for legally required standby systems is permitted to occupy the same raceways with general-purpose wiring.

A) True

B) False

87. A _____ is a location in which combustible dust may be present in the air in quantities sufficient to produce explosive or ignitable mixtures due to abnormal operations.
    A) Class II, Division 1
    B) Class II, Division 2
    C) Class III, Division 1
    D) Class III, Division 2

88. _____ fully-engaged threads create a flame-arresting path for conduit connections in a Class I Division 1 Hazardous (Classified) Location.
    A) 5
    B) 4 ½
    C) 4
    D) 3 ½

89. A covered cable tray is considered a raceway.
    A) True
    B) False

90. Single conductor cables in cable tray shall be _____or larger.
    A) 2/0 AWG
    B) 1/0 AWG
    C) 250 kcmil
    D) 1 AWG

91. Ground rods shall be driven to a depth of 8 feet, except where rock bottom is encountered they may be driven at an oblique angle not to exceed 45° from the vertical, or buried in a trench at least 30 inches deep. For one and two family dwellings, where it is not practicable to achieve an overall maximum primary protector bonding or grounding electrode conductor length of 20 feet for a primary protector for communication systems, a ground rod, driven to a depth of _____ may be used where this rod is bonded to the grounding electrode (system) for the service supply.
    A) 4 feet
    B) 5 feet
    C) 6 feet
    D) 10 feet

92. Where buried, conductive optical fiber cables shall be separated at least _____ from conductors of electric light, power, non-power limited fire alarm circuit conductors, or Class 1 circuit conductors.

A) 12 in.
B) 18 in
C) 24 in.
D) 36 in

93. Where conductors are installed above an electric space-heated ceiling, they are considered to be in an ambient temperature of _____.
A) 30°C
B) 40°C
C) 104°C
D) 50°C

94. Short sections of metal enclosures or raceways used to provide protection or support of cable assemblies _____ be required to be connected to the equipment grounding conductor.
A) shall
B) shall not
C) over 600 volts
D) under 600 volts

95. In Class II, Division 1 Hazardous(Classified) Locations, motors, generators, and other rotating machinery shall be identified for the location or _____.
A) explosionproof
B) dusttight
C) totally enclosed pipe-ventilated
D) dust-ignitionproof

96. An FCC cable consists of _____ or more flat copper conductors placed edge-to-edge and separated and enclosed within an insulating assembly.
A) 2
B) 4
C) 3
D) 5

97. Type TC cables and associated fittings shall be _____.
A) identified
B) listed
C) approved
D) A and B

98. Strut-type channel raceways and accessories shall be _____ for such use.
    A) listed
    B) approved
    C) identified
    D) A and C

99. Schedule 40 PVC is permitted for underground installations, including under driveways, where the required burial depth is met.
    A) True
    B) False

100. Trade size 4″ PVC shall be supported with a maximum spacing of _____ between supports.
    A) 5 feet
    B) 6 feet
    C) 7 feet
    D) 8 feet

101. If a building has a fire sprinkler system installed in accordance with _____ , ENT is permitted within walls, floors, and ceilings, whether exposed or concealed, in buildings exceeding 3 floors above grade.
    A) NFPA 13-2013
    B) a fire finish rating
    C) a thermal barrier
    D) all of these

102. Faceplates provided for snap switches mounted in boxes and other enclosures shall be installed to completely _____ the opening and where flush mounted, _____ against the finished surface.
    A) close – sit
    B) cover – seat
    C) cover – close
    D) B and C

103. Overcurrent devices for photovoltaic source circuits shall be readily accessible.
    A) True
    B) False

104. Threaded covers for enclosures in Hazardous (Classified) Locations must have at least _____ fully engaged threads.
    A) 5
    B) 4½
    C) 4
    D) 3½

105. A nonferrous metal raceway may be used for physical protection of a grounding electrode conductor without bonding at each end.
    A) True
    B) False

106. Three-3/0 uncoated copper conductors installed in parallel have a DC resistance at 75°C. of _____ for a length of 100 feet.
    A) 0.00255 ohms
    B) 0.00456 ohms
    C) 0.00756 ohms
    D) 0.0766 ohms

107. Service equipment at other than dwelling units shall be legibly marked in the field with the _____.
    A) voltage
    B) current
    C) maximum available fault current
    D) minimum working clearance

108. Optical fiber cables installed in vertical runs that penetrate more than one floor without being installed in a raceway shall be type _____.
    A) OFCR-OFCP
    B) OFNR-OFNP
    C) OFC
    D) A and B

109. Type AC and MC Cable without an overall nonmetallic covering are permitted wiring methods in a space over a suspended ceiling used for environmental air-handling purposes.
    A) True
    B) False

110. Type AC cable installed through, or parallel to framing members, shall be protected from physical damage by nails, screws, etc.
   A) True
   B) False

111. The on-site fuel supply where an internal combustion engine is used as the prime mover for the generator of an emergency system shall provide for not less than _____ hours of full operation of the system.
   A) 3
   B) 2
   C) 5
   D) 4

112. A copper main bonding jumper for a service supply consisting of one set of 350 kcmil copper conductors shall be a minimum size of _____.
   A) 1/0
   B) 2/0
   C) 2 AWG
   D) 3/0

113. The minimum working clearance for voltages of 601-1000 volts where there are exposed live parts on both sides of the working space is _____.
   A) 4 feet
   B) 3 feet
   C) 5 feet
   D) 3 ½ feet

114. Where a pool pump motor is supplied through an outlet or wired by a direct connection at 120 volts or 240 volts, single-phase, the circuit shall be provided with _____ protection for persons.
   A) GFCI
   B) AFCI
   C) A and B
   D) A or B

115. Nonmetallic cable ties for the support of cable assemblies in an air-handling space shall be _____ as having low-smoke and heat-release properties.
   A) listed

B) identified
C) A and B
D) A or B

116. Bonding shall be provided where necessary to ensure electrical continuity and the capacity to conduct safely any _____current likely to be imposed.
A) stray
B) conducting
C) fault
D) lightning

117. The _____ shall conduct or witness a test of the complete system upon installation and periodically afterward of an emergency system.
A) qualified person
B) authority having jurisdiction
C) maintenance electrician
D) professional engineer

118. On a 4-wire, 3-phase, Wye connected system, where the major portion of the load consists of nonlinear loads, the neutral conductor is considered to be a current-carrying conductor.
A) True
B) False

119. The ambient temperature correction factor for a conductor with a 75°C. temperature rating at 110° F. is_____.
A) 0.88
B) 0.75
C) 0.82
D) 0.94

120. The minimum size of conductor, where conductors are connected in parallel is _____.
A) 2/0
B) 3/0
C) 1/0
D) 1 AWG

121. The minimum size of a metal wireway that contains four 500 kcmil THW conductors, four 1/0 THHN conductors, and four 300 kcmil conductors is _____.
    A) 8 in. × 8 in.
    B) 6 in. × 6 in.
    C) 4 in. × 4 in.
    D) 12 in. × 12 in.

122. The number of conductors permitted in a box that has two-12/2 W/G NM cables, two duplex receptacles, and two internal cable clamps is _____.
    A) 10 conductors
    B) 8 conductors
    C) 6 conductors
    D) 7 conductors

123. Information Technology Equipment is permitted to be connected to a branch-circuit by a power supply cord that does not exceed _____ in length.
    A) 3 feet
    B) 6 feet
    C) 15 feet
    D) 10 feet

124. The wiring methods permitted under a raised floor in an Information Technology Equipment room include the wiring methods of 300.22(C), and liquidtight flexible metal conduit, and liquidtight flexible nonmetallic conduit.
    A) True
    B) False

125. Industrial machinery shall be legibly marked in the field with the _____.
    A) ampere rating
    B) voltage rating
    C) maximum available short-circuit current
    D) location of disconnecting means

126. Downstream connections between the neutral conductor and the equipment grounding system will produce an _____ current over the equipment grounding conductors.

A) fault
B) objectionable
C) overcurrent
D) isolation

127. Fuses and circuit breakers shall be selected and coordinated to clear a fault without extensive damage to the electrical equipment of the circuit. _____ equipment applied in accordance with their _____ _shall be considered to meet the requirements of this section.
A) identified, recognized
B) identified approved
C) listed, listing
D) A and B

128. The conductors on the load side of the service point are defined as the service conductors.
A) True
B) False

129. Circuit breakers shall _____ all of the ungrounded conductor of the circuit both manually and automatically, unless otherwise permitted.
A) close
B) open
C) connect
D) isolate

130. When normally enclosed live parts are exposed for inspection or servicing, the required working space in a passageway or general open space shall be suitably _____.
A) protected
B) guarded
C) identified
D) marked

131. A flexible cord that is field assembled with 16 AWG conductors is permitted to be protected by a _____ ampere circuit.
A) 20
B) 15
C) 10
D) 25

132. For voltages of 1000 volts, nominal or less, the width of the working space in front of the electrical equipment shall be the width of the equipment or _____, whichever is greater.
    A) 30 in.
    B) 24 in.
    C) 42 in.
    D) 36 in.

133. Circuit breakers used as switches in 120 volt and 277 volt fluorescent lighting circuits shall be listed and marked _____or _____.
    A) SWD
    B) HID
    C) HD
    D) A and B

134. Color coding shall be permitted to identify intrinsically safe conductors where they are colored _____.
    A) light orange
    B) white
    C) black
    D) light blue

135. In dwelling units, 120 volt, single-phase, 15 and 20 ampere branch circuits supplying outlets or devices installed in laundry areas require arc-fault circuit interrupter protection.
    A) True
    B) False

136. Conductors supplied by the secondary side of a single-phase transformer having a 2-wire single-voltage secondary _____ be permitted to be protected by the overcurrent protection on the primary side of the transformer, if this protection does not exceed the value determined by multiplying the secondary conductor ampacity by the secondary-to-primary transformer voltage ratio.
    A) shall
    B) shall not
    C) by special permission
    D) A and C

137. Outlet boxes that are listed for the sole support of a ceiling-suspended paddle fan weighing more than _____ shall be marked for the maximum weight to be supported.
A) 35 pounds
B) 70 pounds
C) 50 pounds
D) 60 pounds

138. Where the DC circuits of solar photovoltaic systems are not covered by PV modules and are embedded in built-up, laminate, or membrane roofing materials, the location shall be clearly marked.
A) True
B) False

139. Conductors of AC and DC circuits, up to 1000 volts, shall be permitted to occupy the same raceway, wiring enclosure, or cable, including circuits operating at 480 volts, 240 volts, and 120 volts, where all of the conductors have insulation ratings equal to at least the maximum circuit voltage applied to any conductor.
A) True
B) False

140. Conduit seals shall be installed within _____ of an explosionproof enclosure where a conduit is 2 inch trade size or larger in a Class I, Division 1 Hazardous (Classified) Location.
A) 18 in.
B) 24 in.
C) 30 in.
D) None of these

141. _____ is a group B material.
A) Propane
B) Ethylene
C) Hydrogen
D) Acetylene

142. The demand load for ten 10kW, eight 16kW, and six 12 kW household electric ranges is _____.
A) 40.95kW
B) 35.10kW
C) 28.50kW
D) 38.65kW

143. The nameplate rating for each electric dryer in a 20 unit apartment building is 4.5 kW. The demand load applied to the service is _____.
    A) 38 kW
    B) 42 kW
    C) 47 kW
    D) 56 kW

144. Conductors that supply a single continuous-duty motor shall have an ampacity of _____of the motor full-load current rating.
    A) 100 percent
    B) 150 percent
    C) 110 percent
    D) 125 percent

145. Continuous-duty motors with a marked service factor of 1.15 shall have a separate overload device rated at no more than _____ of the motor nameplate current rating.
    A) 125 %
    B) 115 %
    C) 130 %
    D) 156 %

146. A 10hp, 460 volt, three-phase, alternating current motor at 80 percent power factor has a full-load current rating of _____.
    A) 14.5 amperes
    B) 16.8 amperes
    C) 17.5 amperes
    D) 20 amperes

147. Ground rods with a ½ inch diameter shall be _____.
    A) copperclad
    B) steel
    C) stainless steel
    D) listed

148. To qualify as a concrete-encased electrode, a bare steel reinforcing rod of not less than ½″ diameter, or a 4 AWG copper conductor must be at least _____ feet long.
    A) 20
    B) 10

C) 30

D) 40

149. Auxiliary or supplemental grounding electrodes, such as ground rods connected to metallic equipment frames, may be used in lieu of an equipment grounding conductor.

A) True

B) False

150. Where a building is supplied by a feeder consisting of 500kcmil copper conductors, a grounding electrode conductor connected to a metal in-ground support structure shall not be smaller than that specified in

_____.

A) Table 250.122

B) Table 250.66

C) Table 250.166

D) Table 310.16(B)(16)

151. Mobile home service equipment shall be located in sight from and not more than _____ feet from the exterior wall of the mobile home.

A) 50 feet

B) 30 feet

C) 10 feet

D) 25 feet

152. _____ shall be provided for critical operations data systems.

A) Surge protection

B) Grounding electrode

C) Disconnecting means

D) None of these

153. Solar photovoltaic system grounding methods must accomplish the equivalent system protection of 250.4(A). This includes a 2-wire array with one _____ grounded conductor.

A) functional

B) resistance

C) reactance

D) None of these

154. The maximum size overcurrent device on the primary side of a 3-phase, 75 kVA transformer, connected Delta/Delta, a primary voltage of 480 volts, and a secondary voltage of 240 volts, and no secondary overcurrent protection is _____.
   A) 125 A
   B) 45 A
   C) 70 A
   D) 80 A

155. The demand factor for 8 recreational vehicle sites is _____.
   A) 50%
   B) 48%
   C) 65%
   D) 55%

156. An individual nonmotor generator arc welder shall have the supply conductors based on no less than the current value determined by multiplying the rated primary current given on the welder nameplate by the factor shown in _____, based on the duty cycle of the welder.
   A) Table 630.31(A)(2)
   B) Table 630.11(A)
   C) Table 310.16(B)(16)
   D) Table 250.122

157. Where one conductor or cable assembly is installed in a conduit or tubing, the percent fill is based on _____.
   A) 31 %
   B) 40 %
   C) 70 %
   D) 53 %

158. When calculating percentage conduit or tubing fill area for a multiconductor cable that has an eliptical cross section, the cable area is based on using the major diameter of the ellipse as a circle diameter.
   A) True
   B) False

159. Where optical fiber cables are installed above a roof, a vertical clearance of not less than _____ feet is required from all points of roofs above which they pass.

A) 8
B) 10
C) 6
D) 12

160. A ground ring shall be installed not less than _____ below the surface of the of the earth.
A) 18 in.
B) 30 in.
C) 42 in.
D) 60 in.

161. Where raceways or cables are exposed to direct sunlight on or above rooftops, the minimum distance above the roof to the bottom of the raceway or cable is to be no less than _____.
A) 1 in.
B) 7/8 in.
C) 3 in.
D) 1/2 in.

162. The ampacities referenced in 310.15 do not take _____ into consideration.
A) voltage –drop
B) conductor fill
C) resistivity
D) A and B

163. A 10hp-3-phase, 460 volt Design B motor has a locked-rotor current of _____ amperes.
A) 46
B) 32
C) 116
D) 81

164. Where a single phase transformer has a 2-wire, single-voltage secondary, and a primary current of less than 2 amperes, the primary overcurrent protection may be _____ of the transformer rated primary current.
A) 300 %
B) 125 %
C) 167 %
D) 250 %

165. A zigzag connected autotransformer used for the purpose of providing a neutral point for grounding purposes _____ be installed on the load side of any system grounding connection, including those made for the service or a separately-derived system.
    A) shall
    B) shall not
    C) is permitted
    D) none of these

166. A general care (Category 2) space in a health care facility is a space in which failure of equipment or system is likely to cause minor injury to patients, staff, or visitors.
    A) True
    B) False

167. The primary overcurrent protective device shall be selected or set to carry indefinitely the sum of the _____ of the fire pump motor and the pressure maintenance pump motor and the fire pump accessory equipment when connected to the power supply.
    A) full-load current
    B) residual current
    C) locked-rotor current
    D) supplementary current

168. Where conduits extend into the bottom of an open switchgear enclosure, the conduits, including their end fittings, shall not rise more than _____ above the bottom of the enclosure.
    A) 5
    B) 6
    C) 3
    D) 4

169. Fluorescent lighting ballasts are required to have _____ protection where installed indoors.
    A) GFCI
    B) supplementary
    C) AFCI
    D) integral thermal

170. The 3 volt-ampere rule from Table 220.12 for dwelling occupancies includes all 15A and 20A receptacles, but not the small appliance and laundry loads.
    A) True
    B) False

171. Coaxial cables may be strapped or taped to the exterior of a raceway as a means of support.
    A) True
    B) False

172. A feeder is protected at 1000 amperes and a set of conductors are tapped to the feeder, with the tap conductors extending 10 feet to a 100 ampere main circuit breaker in a panelboard. Based on the use of THHN copper conductors, what are the smallest size tap conductors permitted?
    A) 3 AWG
    B) 2 AWG
    C) 1 AWG
    D) 1/0 AWG

173. Circuit breakers and other switching devices that are intended to interrupt current under normal conditions in a Class I, Division 2, Hazardous (Classified) Location are required to be installed in a Class I, Division 1 enclosure, unless another means of protection is provided.
    A) True
    B) False

174. Cord sets or devices that have GFCI protection for personnel used for temporary wiring must be _____.
    A) identified
    B) listed
    C) approved
    D) A and B

175. Temporary wiring is permitted to remain in place after the construction is completed for use on renovations.
    A) True
    B) False

176. The emergency system equipment shall be suitable for the _____ at its terminals.
    A) maximum voltage
    B) maximum current
    C) maximum available fault current
    D) maximum ground-fault current

177. Service conductors passing over track rails of railroads at voltages up to 600 volts must have a clearance of _____.
    A) 24 feet
    B) 24½ feet
    C) 18 feet
    D) 18½ feet

178. Live parts operating at 50 to 1000 volts shall be_____by approved enclosures or protected by other means, such as by elevation.
    A) guarded
    B) enclosed
    C) insulated
    D) identified

179. A NEMA 3R enclosure protects against rain, snow, and _____, and may be marked rainproof.
    A) corrosive agents
    B) hose down
    C) windblown dust
    D) sleet

180. 2-wire DC systems that supply premises wiring operating at greater than 60 volts, but not greater than 300 volts shall be _____.
    A) Grounded
    B) Isolated
    C) GFCI protected
    D) AFCI protected

181. The minimum clear working space for live parts of equipment operating at 300 volts-to-ground is _____ feet.
    A) 4 feet
    B) 5 feet
    C) 3 feet
    D) 3 ½ feet

182. Working spaces about battery systems shall be in accordance with 110.26. The working space is measured from the edge of the _____.
    A) batteries
    B) battery cabinet
    C) racks or trays
    D) B and C

183. A disconnecting means shall be provided for all ungrounded conductors from a stationary battery system with a voltage over _____DC.
    A) 30 volts
    B) 60 volts
    C) 150 volts
    D) 100 volts

184. Overcurrent protection is permitted _____to the storage battery terminals in an unclassified location.
    A) in sight from
    B) within 6 feet
    C) as close as practicable
    D) none of these

185. Circuit breakers have a minimum interrupting rating of _____ amperes.
    A) 10,000
    B) 5,000
    C) 22,000
    D) 1,000

186. A metal box with a dimension of 4"x1½" (square) has a minimum volume of _____ cubic inches and may have a maximum of _____ 12 AWG conductors.
    A) 21 and 9
    B) 18 and 8
    C) 21.5 and 9
    D) 30.3 and 13

187. A locknut that is intended to perform a mechanical function as opposed to an electrical function is described as a _____.
    A) accessory
    B) device
    C) part
    D) fitting

188. If air-conditioning and refrigerating equipment does not have a hermetic refrigerant motor-compressor, the rules of Article _____ apply.
    A) 430
    B) 440
    C) 422
    D) 424

189. The supply for sensitive electronic equipment is a separately-derived system operating at _____ line-to-line and _____ to ground.
    A) 120 volts – 60 volts
    B) 240 volts – 120 volts
    C) 60 volts – 30 volts
    D) 120 volts – 30 volts

190. The transformer secondary center tap of the 60/120 volt, 3-wire system is not required to be grounded when supplying sensitive electronic equipment.
    A) True
    B) False

191. Voltage-drop on any branch circuit supplying sensitive electronic equipment is limited to _____, and the combined voltage-drop of feeder and branch circuit conductors shall not exceed _____.
    A) 3 % – 5 %
    B) 1.5 % – 2.5 %
    C) 1 % – 3 %
    D) 2.5 % 4.5 %

192. Flexible cord listed for _____ and terminated with listed dusttight cord connectors is permitted in Class II, Division 1 Locations.
    A) Extra-hard useage
    B) hard-useage
    C) thermoset type
    D) A and B

193. Outlet boxes may be secured to suspended-ceiling framing members by mechanical means, such as _____.
    A) bolts
    B) rivets
    C) screws

D) any of these

194. Permanently installed swimming pools include those constructed in-ground, partially in-ground, or above-ground capable of holding water in a depth greater than _____ inches.
A) 48
B) 42
C) 36
D) 24

195. Access to electrical equipment shall not be denied by an accumulation of remote-control, signaling, or power-limited wires and cables that prevents the removal of panels, including suspended ceiling panels.
A) True
B) False

196. The maximum voltage for one and two family dwellings from wind turbine output circuits is _____ .
A) 240 volts
B) 300 volts
C) 480 volts
D) 600 volts

197. Cable trays used for the support of nonconductive optical fiber cables and Class 2 and Class 3 Remote Control Signaling and Power Limited Circuits shall be electrically continuous through approved connections or the use of a bonding jumper.
A) True
B) False

198. The grounding electrode conductor for an antenna mast shall not be smaller than 6 AWG copper.
A) True
B) False

199. A Supply-Side Bonding Jumper for alternating current systems is sized in accordance with _____.
A) 250.66
B) 250.122
C) 250.102(C)(1)
D) 250.166

200. For circuits of over 250 volts to ground, bonding through the use of 2 locknuts on rigid metal conduit or intermediate metal conduit on clean knockouts, or a box or enclosure with concentric or eccentric knockouts listed to provide a reliable bonding connection is permitted.
    A) True
    B) False

Answer Key

1. C	-	356.24
2. A	-	240.24(F)
3. A	-	402.6
4. A	-	725.25
5. A	-	810.21(J)
6. A	-	250.32 (A), Exception
7. A	-	110.26 (E) (1), (a)
8. B	-	701. 3 (A), (B)
9. A	-	Table 300.5 Column 4
10. C	-	445.1
11. B	-	230.95
12. B	-	Table 110.28 –Article 100-Definition-Rainproof
13. A	-	314.28(B)
14. B	-	215.6
15. B	-	800.53 – 820.44 (E)(3)
16. A	-	517.13(B) (1), Exception No.2
17. B	-	690.10 – 710.15(E) 408.36(D)
18. B	-	695.6(G)
19. D	-	690.43(B)
20. A	-	690.41 (B)(2),(1),(2)
21. B	-	502.15
22. B	-	501.125(A)
23. B	-	350.30(A), Exception No. 1
24. B	-	340.10 (4) –Article 334-Parts II and III
25. B	-	354.12(2)
26. A	-	358.10(A)(1)
27. B	-	338.12(B)(1)
28. B	-	344.42(B) – 342.42(B)
29. D	-	300.6 – 342.10(D) – 344.10(D)
30. D	-	250.80 – 250.86 – 250.96(A) – 314.4
31. A	-	250.112(I)

32. A - 230.82 (8)
33. A - 285.26
34. A - 285.6
35. B - 285.7
36. D - 285.12 – 250.4(A) Informational Note 1.
37. D - 285.23(B) – Article 100-Definition
38. A - 280.4(A)(2) –Article 100-Definition
39. B - 504.50 (A)
40. A - 225.33 (A)
41. A - 692.6
42. B - 110.14(C)(1),(b)
43. C - 250.92(B)(3)
44. A - 820.133(A)(1)(a)(3)
45. C - Table 310.106(A)
46. B - Table 250.66
47. B - 250.66 (B)
48. B - 250.50 – 250.58 - 800.100(D)
49. A - Table 250.66 – Note 2
50. C - 250.53(A)(2), Exception
51. B - 250.52(B)(3)
52. B - 250.34(A)
53. D - 250.4(A)(1)
54. A - 250.60 – 250.106 – 250.50 – 250.58
55. A - 690.47 (A)
56. D - 250.30 (A)(1)
57. A - Article 100 – Definition – Separately-Derived System
58. A - 250.122(B)
59. B - 690.45
60. C - 250.118(8) – 320.108 – 250.4(A)(5) – 250.4(B)(4)
61. A - 220.55
62. C - Table 220.54
63. C - Table 220.44
64. C - 250.53(G)
65. A - 430.102 (B)(2), Exception (a)
66. A - Article 100 – Definition – Neutral point
67. B - 90.1(B)
68. D - 314.3
69. B - 625.2 – Definition-Wireless-Power Transfer Equipment
         Article 100-Separately Derived System
70. A - Table 250.66

71. C - 250.53(B)
72. A - 250.53(A)(3), Informational Note
73. C - 625.52(A)
74. B - 250.54
75. D - Article 100-Definition-Remote Control Circuit
76. C - 285.7
77. B - 250.122(F)(2),(b)
78. A - 518.4(A)
79. A - 700.5(C)
80. B - 700.5(E)
81. A - Table 402.5
82. B - 376.56(A)
83. B - 680.26(B)
84. B - 555.24
85. A - 645.14
86. A - 701.10
87. B - 500.5(C)(2)
88. A - 500.8(E)(1)
89. B - Article 100 – Definition – Raceway
           392.2-Definition – Cable Tray System
90. B - 392.10 (B)(1)(a)
91. B - 800.100 (A)(4) Exception – 800.100(B)(3)(2)
92. A - 770.47 (B)
93. D - 424.36
94. B - 250.86, Exception 2
95. C - 502.125 (A)(1),(2)
96. C - 324.2 – Definitions
97. B - 336.6
98. D - 384.6
99. A - 352.10(G) – 300.5 – 300.50
100. C - Table 352.30
101. A - 362.10 (2), Exception
102. B - 404.9(A)
103. B - 690.9(C)
104. B - 500.8(E)(1), Exception
105. A - 250.64(E)(1)
106. A - Table 8 – Chapter 9

$$\frac{0.0766\,ohms \times .1\,(100\,feet)}{0.00766\,ohms}$$

$$\frac{0.00766\,ohms}{3\,conductors} = 0.00255\,ohms$$

107. C - 110.24(A)
108. D - 770.113(D)(1)
109. A - 300.22(C)(1)
110. A - 320.17 – 300.4(A),(C),(D)
111. B - 700.12(B)(2)
112. C - Table 250.102(C)(1)
113. C - Table 110.26(A)(1)
114. A - 680.21(C)
115. A - 300.22(C)(1)
116. C - 250.90
117. B - 700.3(A),(B)
118. A - 310.15(B)(5),(c)
119. C - Table 310.15(B)(2),(a)
120. C - 310.10(H)(1)
121. B - 376.22(A) – Table 5-Chapter 9
         Total area of conductors
         500kcmil –THW 0.7901 square inches
              $\times$ 4
         3.1604 square inches
         1/0- THHN   0.1855 square inches
                  $\times$ 4
                    0.7420 square inches
         300kcmil-THHN – 0.4608 square inches
                  $\times$ 4
                    1.8432 square inches
         3.1604 square inches
         0.7420 square inches
         1.8432 square inches
         5.7456 square inches –Total
         6″ × 6″ wireway- 36 square inches
                  $\times$ .2(20% fill)-376.22(A)
                    7.2 square inches
122. A - 314.16(B)(1), (2), (4), (5)
123. C - 645.5(B) (1)
124. A - 645.5(E)(1),(11),(12)
125. C - 670.5(2)
126. B - 250.6(A),(B)

127. C - 110.10
128. A - Article 100 – Definition – Service Conductors
129. B - 240.15(B)
130. B - 110.26(B)
131. A - 240.5(B)(4)
132. A - 110.26(A)(2)
133. D - 240.83(D)
134. D - 504.80(C)
135. A - 210.12(A)
136. A - 240.4(F)
137. A - 314.27(C)
138. A - 690.31(G)(1)
139. A - 300.3(C)(1)
140. A - Table 501.1 – 501.15(A)(1),(2)
141. C - 500.6(A)(2), Informational Note
142. A - Table 220.55 – Note 2

    10 – 10 kW ranges = 10 ranges × 12 kW = 120 kW
    8 – 16 kW ranges = 8 ranges × 16 kW = 128 kW
    6 – 12 kW ranges = 6 ranges × 12 kW = 72 kW
                                            320 kW

$$\frac{320 kW}{24 \text{ Ranges}} = 13.33 \text{ kW}$$

    13.33 kW
    −12.00 kW
      1.33kW (drop fraction less than .50kW)
    1.00kW × .05=5%
    Required demand increase = 5% (1.05) Column C of Table
    220.55 for 24 Ranges = 39kW
    39 kW
    × 1.05 kW

40.95 kW Total Demand

143. A - 220.54 – Table 220.54
    220.54 Requires a minimum of 5000 watts for each dryer × 20
    dryers = 100 kW Table 220.54 – demand factor for 20 dryers-
    47% minus 1% for each dryer exceeding 11 = 47 − 9 = 38%

    100,000 *watts*
    ×   .38
    _____
    38,000 *watts, or* 38 *kW*

144. D - 430.22

145. A   -   430.32 (A)(1)
146. C   -   Table 430.250 – Note

$$\frac{14 \ amperes}{\times 1.25} \quad \frac{-Table\,430.150}{Note - 80\%\,PF}$$
$$17.5 \ amperes$$

147. D   -   250.52 (A)(5),(b)
148. A   -   250.52(A)(3)
149. B   -   250.54
150. B   -   250.52(A)(2)
151. B   -   550.32(A)
152. A   -   645.18
153. A   -   690.41(A)(1)
154. A   -   (Table 450.3(B) - Note 1)

$$\frac{75.000\,VA}{480\,V. \times 1.732} = 90.21\,A \quad \begin{array}{c} 90.21\,A \\ \times 1.25 \\ \hline 112.76\,A \end{array}$$

Next standard size of overcurrent device is 125 amperes (240.6(A))

155. D   -   Table 551.73(A)
156. B   -   630.11(A)
157. D   -   Table 1 – Chapter 9
158. A   -   Chapter 9- Notes to Tables – Note (9)
159. A   -   770.44(B)
160. B   -   250.53(F)
161. B   -   310.15(B)(3),(c)
162. A   -   310.15 (A)(1) Informational Note No. 1
163. D   -   Table 430.251(B)
164. A   -   Table 450.3(B)
165. B   -   450.5
166. A   -   517.2 – Definition – Patient Care Space
167. C   -   695.5(B)
168. C   -   408.5
169. D   -   410.130(E)(1)
170. A   -   220.52(A),(B)
171. B   -   820.133(B)
172. C   -   240.21(B)(1) – Table 310.15(B)(16)-110.14(C)(1)(a)(1)
173. A   -   501.115 (B)(1)
174. B   -   590.6(A)(1)

175. B   -   590.3(D)
176. C   -   700.4(A)
177. B   -   230.24(B)(5)
178. A   -   110.27(A)
179. D   -   Table 110.28
180. A   -   250.162(A)
181. A   -   110.26(A)(1) – Table 110.26(A)(1)
182. D   -   480.10(C)
183. B   -   480.7(A)
184. C   -   240.21(H)
185. B   -   240.83(C)
186. A   -   Table 314.16(A)
187. D   -   Article 100- Definition - Fitting
188. C   -   422.3
189. A   -   647.1
190. B   -   647.6(A)
191. B   -   647.4(D)
192. A   -   502.10(A)(2),(5)
193. D   -   314.23(D)(1)
194. B   -   680.2 – Definition – Permanently Installed
             Swimming, Wading, Immersion, and Therapeutic Pools
195. A   -   725.21
196. D   -   694.10(A)
197. A   -   392.60(A)
198. B   -   810.21(H)
199. C   -   Table 250.102(C)(1)
200. A   -   250.97 Exception and 250.97, Exception 2

Index

About the Author

It started on August 3, 1964 as I started to work in the electrical trade through the auspices of Local 52, IBEW in Newark, N.J. Seven months later, I became a member of this Local, which was merged with Local 164 (Paramus, N.J.) in 2000.

I served my apprenticeship until March 1, 1969, at which time I became a Journeyman-Wireman. I gained a wealth of knowledge and experience during this period and the years that followed.

In March of 1978, I developed an association with a training/consulting company in Trenton, N.J. This company had a request from a client in Philadelphia to present an electrical training course for their maintenance personnel. The class was scheduled for four weeks. I was asked to conduct this course, and despite not having any teaching experience, I decided to become an instructor. Fortunately, the class went well, and the training was extended for another four weeks. In the meantime, this training/consulting company offered me a permanent position.

In September of 1978, I started to offer courses on the topic of the National Electrical Code. The key was to find a method of instruction that would benefit the students by keeping their attention during the three day course period. And, to foster an interest in this complex document that would serve them well beyond our brief time together. In later years, I developed and presented courses on the topics of Grounding Electrical Distribution Systems, Designing Overcurrent Protection, Electrical Systems In Hazardous (Classified) Locations, and Electrical Equipment Maintenance. These courses were offered by my company, Electrical Design Institute, and several universities, including the University of Wisconsin, George Washington University, North Carolina State University, the University of Toledo, and the University of Alabama.

In 2021, I authored books entitled, The NEC and You, Perfect Together, Grounding Electrical Distribution Systems, and Designing Overcurrent Protection, NEC Article 240 and Beyond. These books are published by River Publishers.

Gregory P. Bierals
May 24, 2021